PIEZOELECTRIC ACTUATORS: VECTOR CONTROL METHOD

PIEZOELECTRIC ACTUATORS: VECTOR CONTROL METHOD

Basic, Modeling and Mechatronic Design of Ultrasonic Devices

FRÉDÉRIC GIRAUD
CHRISTOPHE GIRAUD-AUDINE

ELSEVIER
Butterworth-Heinemann
An imprint of Elsevier

Butterworth-Heinemann is an imprint of Elsevier
The Boulevard, Langford Lane, Kidlington, Oxford OX5 1GB, United Kingdom
50 Hampshire Street, 5th Floor, Cambridge, MA 02139, United States

Library of Congress Cataloging-in-Publication Data
A catalog record for this book is available from the Library of Congress

British Library Cataloguing-in-Publication Data
A catalogue record for this book is available from the British Library

ISBN: 978-0-12-814186-1

For information on all Butterworth-Heinemann publications
visit our website at https://www.elsevier.com/books-and-journals

Publisher: Mara Conner
Acquisition Editor: Sonnini Yura
Editorial Project Manager: Leticia Lima
Production Project Manager: Sruthi Satheesh
Designer: Mark Rogers

Typeset by VTeX

Contents

1. Introduction **1**

1.1. What is a piezoelectric actuator? 2
1.2. Applications 19
1.3. Variability of the operating point of piezoelectric actuators 23
1.4. Control of piezoelectric actuators 32

2. Dynamic modeling of piezoelectric media **43**

2.1. Introduction 44
2.2. Notations 46
2.3. Elements of elasticity 50
2.4. Dynamic equation of an elastic body 60
2.5. Dynamic of piezoelectric bodies 74
2.6. Practical considerations 85

3. Modeling in a rotating reference frame **95**

3.1. Causal modeling of a piezoelectric transducer 96
3.2. Introduction to the rotating reference frame 103
3.3. Modeling in the rotating reference frame 106
3.4. Modeling a traveling wave in a linear beam 114

4. Control in the rotating reference frame **137**

4.1. Introduction 138
4.2. Control schemes 138
4.3. Controller design 145
4.4. Observers 157

5. Applications **161**

5.1. Introduction 161
5.2. Control of a Langevin transducer 162
5.3. Control of a traveling wave in a finite beam 165
5.4. Control of a traveling wave ultrasonic motor 170

6. Practical implementation **177**

6.1. Introduction 177
6.2. Embedding VCM into a microcontroller 178
6.3. Vibration sensor, amplifier and calibration 187
6.4. Simplification for $V_q = 0$ and a switching converter 195

Bibliography *201*

Index *205*

CHAPTER ONE

Introduction

Contents

1.1. What is a piezoelectric actuator? 2
  1.1.1 Basic principle of a piezoelectric actuator 2
  1.1.2 Advanced piezoelectric actuators 5
    1.1.2.1 Multilayer actuators 6
    1.1.2.2 Benders 8
    1.1.2.3 Amplified actuators 9
  1.1.3 Resonating actuators 10
  1.1.4 Piezoelectric motors 14
    1.1.4.1 PAD motor 16
    1.1.4.2 Ultrasonic motors 16
1.2. Applications 19
  1.2.1 Precise positioning 19
  1.2.2 Industrial processes 21
  1.2.3 Haptic stimulators 21
1.3. Variability of the operating point of piezoelectric actuators 23
  1.3.1 Effect of material's nonlinearity 24
    1.3.1.1 Increased power losses 24
    1.3.1.2 Saturation effect in an ultrasonic actuator 25
    1.3.1.3 Resonance frequency shift 26
  1.3.2 Effect of the temperature 29
  1.3.3 Effect of the mechanical load 30
1.4. Control of piezoelectric actuators 32
  1.4.1 Control basics 32
  1.4.2 Position control of piezoelectric actuators 38
  1.4.3 Vibration control of ultrasonic actuators 39

Abstract

This chapter presents a preliminary overview of the state-of-the art in the field of piezoelectric actuators. Starting from the material, and describing its manufacturing process, we present several ways to exploit the energy conversion in order to obtain the displacement of a mechanical load. The basic principles of some widely used piezoelectric actuators and motors are presented, and illustrated with typical applications. The chapter also discusses the variability of the operating point of the piezoelectric actuators, and explains why a closed-loop control is needed. The chapter ends with the basics of the control theory, which are used in later chapters to design controllers.

Piezoelectric Actuators: Vector Control Method
https://doi.org/10.1016/B978-0-12-814186-1.00005-3

Keywords

piezoelectric actuator, equivalent circuit, multilayer actuator, bender, Langevin transducer, ultrasonic motor, closed-loop control

1.1. What is a piezoelectric actuator?

1.1.1 Basic principle of a piezoelectric actuator

An actuator converts electrical energy into mechanical energy. The piezoelectric actuators rely on piezoelectric materials which are considered as *smart materials* because they have the property to change their dimension when an electrical field is applied on them. Hence, unlike the electromagnetic actuators which produce forces from a distance onto a mechanical load, piezoelectric actuators exploit the strains and stresses generated inside the matter by the inverse[1] piezoelectric effect. In some applications, they can drive the mechanical load directly without gears, to obtain very fine and precise displacements or to move a load with a high force. Hence we find a piezoelectric actuator in an atomic force microscope to keep the needle close to the specimen, or in some diesel engines to open and close the injection valves.

A few crystals can be piezoelectric; in particular, their lattice should not have a center of symmetry. For these materials, which are *dielectric* also, charges appear on their surfaces when they are pressed, due to ionic polarization which results from a relative displacement between positive and negative ions in the crystal's lattice. However, most of the applications described in this book use piezoelectric ceramics which exhibit larger strains and stresses. These materials are *ferroelectric*, which means that they exhibit spontaneous electric dipoles resulting in a polarization P represented by an arrow in Fig. 1.1 (Ballas, 2007). In industrial processes, to create an elementary actuator which consists of a bulk piece of material, one can heat a mix of powders of adequately chosen materials at a temperature high enough to allow the powder grains to fuse with each other, but low enough not to melt the mix. This process is called sintering and the recipe of the mix as well as the temperature profile are secrets jealously guarded by the manufacturers. The so-called PZT ceramic, for example, is very popular, and consists of powders of lead, zircon and titanium alloys. Some dopant materials can be added, like barium, lanthanum, niobium, in order

[1] Piezoelectricity is a reversible process, and electric charges can be produced from a mechanical deformation, which actually is the direct piezoelectric effect.

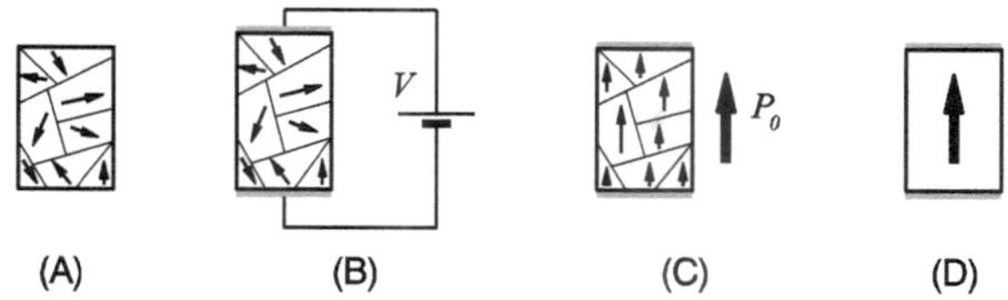

Figure 1.1 A bulk piece of piezoelectric material. After sintering, the poling direction of each domain is randomly distributed (A). A high voltage is applied across the sample, aligning the domains (B). After poling, an electrical polarization P_0 remains because all the domains have the same poling direction (C). A representation of a piezoelectric actuator is shown in (D).

to precisely tune the properties of the final ceramic. The material can be a *hard* piezoelectric, showing a wide linear drive region, but with small strain magnitude, or a *soft* piezoelectric having a high field induced strain with a relatively large hysteresis.

Two electrodes consisting of silver painting are then spread on two faces of the piece using the silk screen method. At this point the material is not piezoelectric. Indeed, the matter is organized in domains, each one having a randomly distributed poling direction. To align them, a high electric voltage (greater than 3 kV/mm) is applied across the two electrodes; during this process, which is named *electric poling*, the sample is placed in a hot bath of oil, the rotation of the domain is facilitated by the temperature increase and the oil has a good dielectric strength. After cooling and cleaning, the sample is ready for use.

When applying a voltage V across the two electrodes, the actuator's length increases, and, ideally, the relationship between stroke and voltage is linear, as illustrated Fig. 1.2 (A). In practice, however, *hysteresis* (the stroke depends on voltage and history), *zero voltage drift* (a residual strain exists at no voltage) and *creep* (strain slowly changes while the voltage is maintained constant due to depoling) occur. These effects are illustrated Figs. 1.2 (B)–(C).

In fact, the actuator not only expands in length, but also shrinks in the lateral directions. Hence, the deformation of the bulk piece occurs in its three dimensions. Depending on the sample's geometry, the lateral displacements can be negligible, or dominant. Table 1.1 shows the displacement in the longitudinal direction (along the polarization) and in the transverse direction (perpendicular to the polarization) for two geometries. This table shows that by choosing the size of the bulk piece of piezoelectric material, it is possible to promote the *longitudinal mode* where the stroke in the lon-

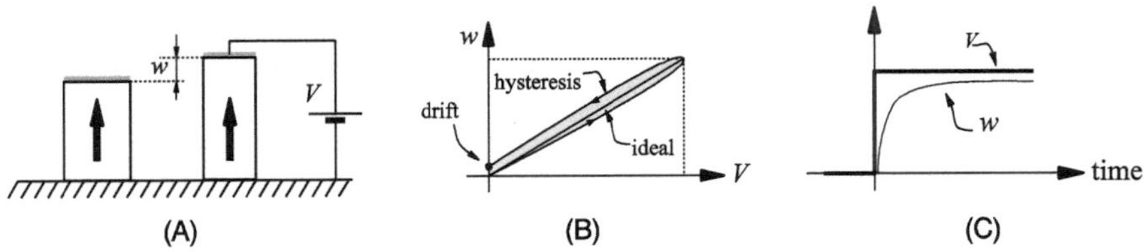

Figure 1.2 When a voltage *V* is applied across the actuator's electrodes (A), the stroke *w* increases, but hysteresis (B) or creep (C) can occur.

Table 1.1 Elongation of a piezoelectric actuator for the longitudinal and transverse direction, supply voltage of 1 kV and typical piezoelectric material.

Actuator type		Size	Longitudinal (z)	Transverse (x)
Rod	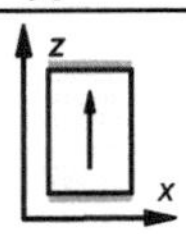	$1 \times 1 \times 2$ cm^3	428 nm	−120 nm
Plate	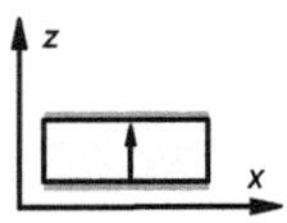	$3 \times 1 \times 0.2$ cm^3	320 nm	−840 nm

gitudinal direction is dominant, or the *transverse mode* where the stroke in the transverse direction is dominant.

With the longitudinal and transverse modes, the voltage is applied in the direction of the remanent polarization. A third coupling mode exists. The *shear mode* produces a rotation of two faces of the material, without changing the length in the other directions. The shear mode is obtained when the poling direction is made in the length of the material, while the voltage is applied perpendicularly to the remanent polarization. As a consequence, a first set of electrodes are painted on two lateral faces of the actuator for the poling process, and then removed. Then, two other electrodes are painted for the electrical voltage to be applied. Finally, the coupling modes are depicted; see Fig. 1.3.

For all coupling modes, the piezoelectric actuators are characterized by the very small displacements they produce. Typically, a $1 \times 1 \times 1.5$ cm^3 piece of material supplied with a 1 kV voltage expands by a few hundreds of nanometer. On the other hand, they can expand while pushing with a high force. For instance, the same sample of piezoelectric material can lift up a load of 200 kg. Hence, they are classified as high force–low speed actuators. From an electrical point of view, piezoelectric actuators are capacitive, and

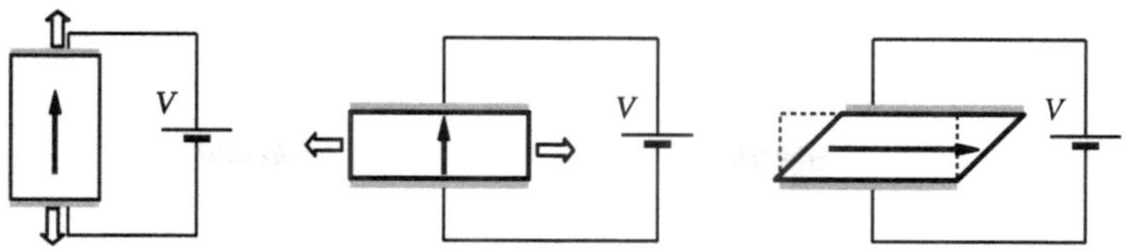

Figure 1.3 Coupling modes of a piezoelectric actuator (→ poling direction, ⇒ dominant displacement): (A) longitudinal, (B) transverse, and (C) shear.

an electrical equivalent circuit is given in Fig. 1.4. In this figure, v is the voltage across the two electrodes, and i_m is a *motional current*. For now, we suppose that i_m is proportional to the displacement speed of the actuator, but this will be demonstrated in Chap. 2.

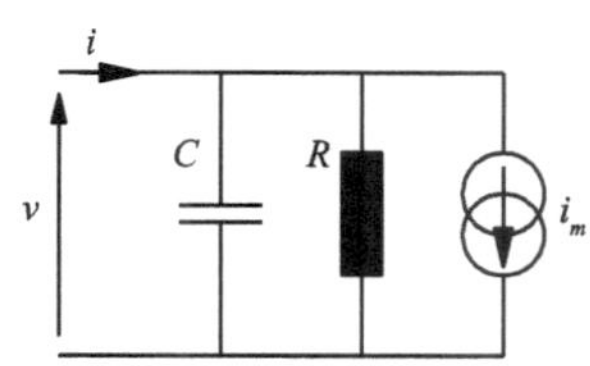

Figure 1.4 An electrical equivalent circuit of a piezoelectric actuator.

The piezoelectric actuator can then be compared with an electromagnetic one, as shown in Table 1.2. It appears that in the equivalent circuit of an electromagnetic actuator, the inductor can be replaced by a capacitor, the voltage source can be replaced by a current source, and the general organization can be changed from serial to parallel, in order to obtain an equivalent circuit of a piezoelectric actuator. Therefore, they can be considered as dual to each other.

1.1.2 Advanced piezoelectric actuators

One disadvantage of the bulk piezoelectric actuators is that they need a high voltage to operate. To reduce the voltage requirement for the same displacement, manufacturers have devised many ways. In this chapter, we describe three types of advanced piezoelectric actuator:

1. Multilayer (or *stack*) actuator,
2. Bender,
3. Amplified actuator.

Table 1.2 Comparison of an electromagnetic actuator with a piezoelectric one.

	Electromagnetic	**Piezoelectric**
Basic actuator	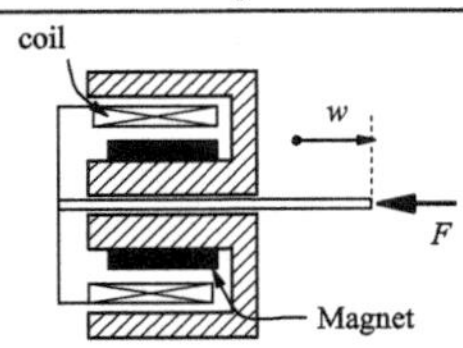	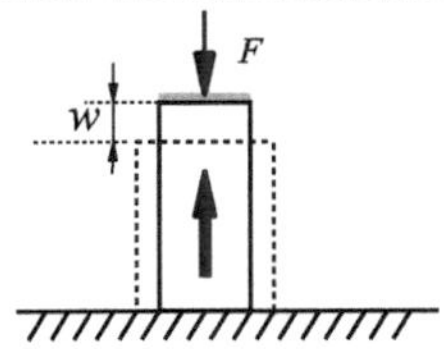
Equivalent circuit	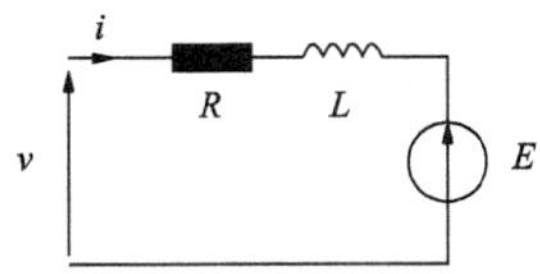	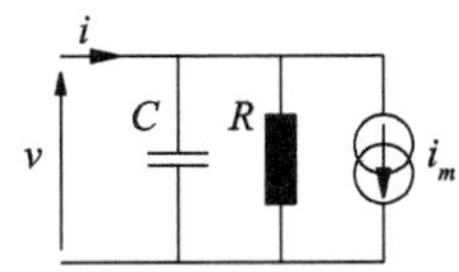
Energy conversion	$F = Ki$ and $E = K\frac{dw}{dt}$	$i_m = N\frac{dw}{dt}$ and $F_p = Nv$
Key features	• large stroke low force • low stiffness, remote action	• small stroke high force • high stiffness, contact with the load

1.1.2.1 Multilayer actuators

Multilayer or stack actuators consist of several plates of the same size, stacked one on top of the other, with alternate poling direction, and electrically connected in parallel. Hence, when a voltage is supplied to the electrodes of the plates, each elementary elongation of the plates is accumulated, leading to a larger stroke. The principle of a multilayer actuator is presented in Fig. 1.5 (A).

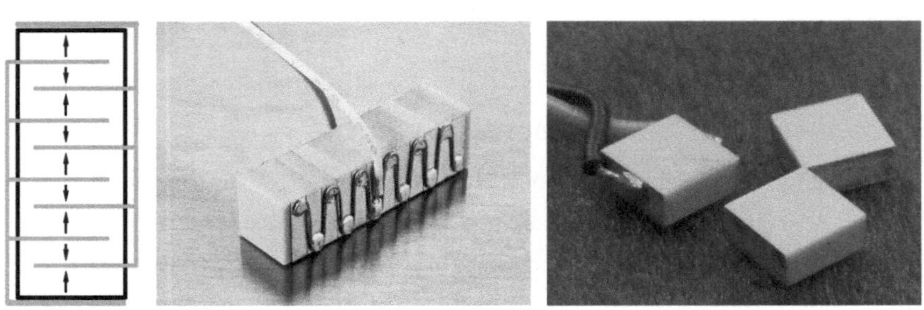

(A) (B) (C)

Figure 1.5 Multilayer piezoelectric actuator: (A) principle, (B) thick plates (www.noliac.com), and (C) thin layers.

Table 1.3 Characteristics of multilayer piezoelectric actuators for different section and length of the same area (25 mm^2). Source www.noliac.com.

Shape	Section	Length	Max elongation	Max force
Square	5 mm × 5 mm	6 mm	6 μm	1050 N
Square	5 mm × 5 mm	10 mm	12 μm	1050 N
Ring	$\Phi_{out} = 6$ mm, $\Phi_{in} = 2$ mm	10 mm	13.2 μm	1060 N

The plates used for such a purpose can be obtained from the sintering process with the raw material in the form of powders (see Fig. 1.5 (B)). But to obtain thinner plates and thus stack more in a given volume, manufacturers use the tape casting method instead. Tape casting consists in mixing the raw material with solvent in order to obtain a liquid. Then, the liquid is poured on a tape and dried.[2] After drying, electrodes are painted on the surface, and the layers are stacked together. A lateral electrode is added to connect the intermediate electrodes together. The process ends with a debinding (which removes the additives), sintering and poling. Thicknesses as low as 40 μm with as much as several hundreds of layers can be obtained with this method (see Fig. 1.5 (C)).

Multilayer actuators have advantages and disadvantages. The main advantage is the deformation-to-voltage ratio, which is larger compared with a bulk actuator of the same size: it is multiplied by the number of layers. Moreover, for two actuators with the same layers thickness, the longer the actuator, the more it will deform for the same voltage. In addition, the force capacity of the actuator is directly related to its section. These features are summarized in Table 1.3, where we give the main characteristics of three actuators from the same manufacturer, but with different shapes.

The disadvantages are twofold. First, by decreasing the voltage, the current requirement increases: for the same dynamic displacement profile, the current is multiplied by the number of layers. Indeed, the capacitance C of an electrical equivalent circuit is equal to the sum of all the small capacitances of each layer because they are connected in parallel. This high current thermally stresses the electrodes, and in particular, the lateral ones are stretched during the deformation. As a consequence, it is necessary to limit the current supplied to the actuator and, as a result, its dynamic. Moreover, because the actuator is not a bulk piece of material, but rather a composite one, it is important to guarantee that the stress profile inside the

[2] The device which casts and adjusts the height of the liquid on the tape is a doctor blade.

component does not lead to inner electrodes' detachment. It is then important to maintain quasi-static operations of the actuator; for that purpose, the manufacturer always specifies a maximum operating frequency, below which it is recommended to operate.

1.1.2.2 Benders

Benders use the amplification effect already known in a bimetal strip. When two plates bonded to each other expand by a different length, the equilibrium position is obtained when the assembly is bent. In the case of a piezoelectric bender, a realization consists in gluing a piezoelectric actuator working in transverse mode onto a thin passive plate, like copper, for example. When applying the voltage, the piezoelectric layer expands (or contracts) while the passive layer's length remains unchanged. This change in the strain of the two layers forces the actuator to bend. This principle is illustrated Fig. 1.6. The main advantage is that a very small difference in the deformation in the material leads to a large stroke of the actuator.

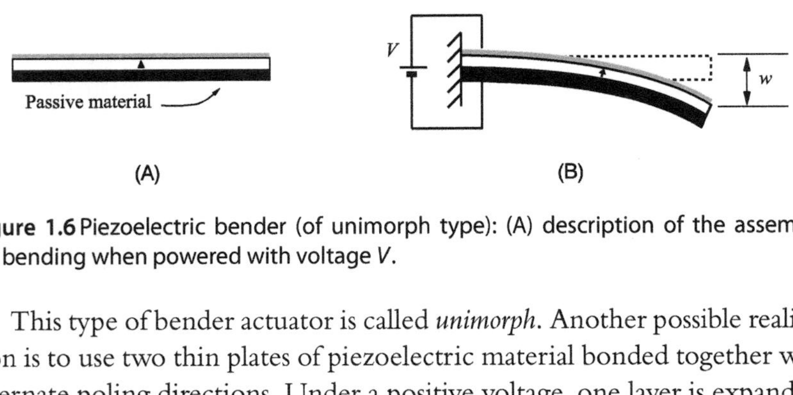

Figure 1.6 Piezoelectric bender (of unimorph type): (A) description of the assembly, (B) bending when powered with voltage *V*.

This type of bender actuator is called *unimorph*. Another possible realization is to use two thin plates of piezoelectric material bonded together with alternate poling directions. Under a positive voltage, one layer is expanding while the other is shrinking, forcing the assembly to bend. This type of bender is called *bimorph*. To allow positive or negative stroke, the bimorph can have a buried electrode which is connected to a variable voltage V, while the top and bottom electrodes are connected to $+V_M$ and $-V_M$, respectively. If V is positive, then the voltage across the top layer is smaller than the voltage across the bottom layer. Hence, the top layer expands less than the bottom layer, forcing the actuator to bend in the downward direction. Conversely, if voltage V is negative, then the actuator bends in the upward direction (Fig. 1.7).

With piezoelectric benders, stroke can be very large, in the range of millimeters. However, the device is not stiff, and its force is limited. For

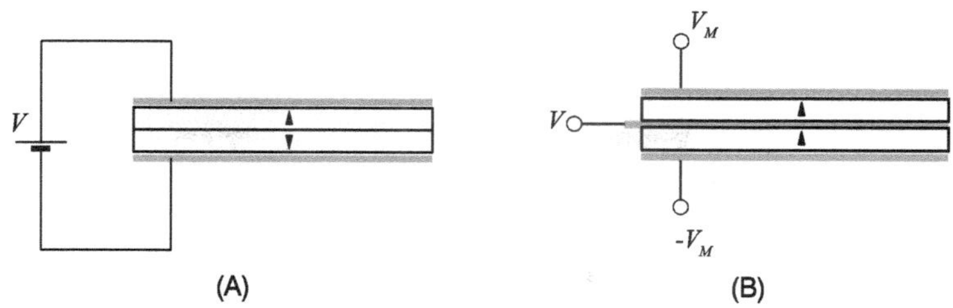

Figure 1.7 Piezoelectric bimorphs: (A) positive stroke only, (B) positive and negative stroke.

example, a $32 \times 8 \times 1$ mm^3 actuator can have a free stroke of 300 µm, but a maximum force of only 2 N. The stiffness is usually less than 10 N/mm. Hence, designing a piezoelectric bender results in a trade-off between the force capability and stroke, or the stroke and stiffness of the device.

1.1.2.3 Amplified actuators

To increase the stroke of the piezoelectric actuators, some manufacturers associate them with a lever arm. A lever arm is often used in mechanisms to increase their force capability: a stiff bar is connected to a pivot (or hinge), named the fulcrum; the load and actuator are positioned at different distances from the pivot. This way, the force applied by the bar to the load can be increased (resp. decreased) with respect to the force of the actuator on the bar. In turn, to maintain energy conservation, the displacement of the actuator can be increased (resp. decreased) with respect to the displacement of the load, and the product *force* × *displacement* is the same for the load and actuator.

Under ideal conditions (no friction in the pivot and stiff bar) the ratio of the stroke is equal to the ratio of the distances of the load to the pivot and of the actuator to the pivot. A ratio higher than 1 means that the force applied to the load is increased, leading to a lower displacement at the load than at the actuator. With a lever arm, a ratio of 10 can be achieved. A higher ratio results in bending of the arm or oversized mechanisms. Fig. 1.8 (A) presents an example of an amplified actuator with a ratio of 3. The pivot is an elastic joint; this ensures reduced backlash in the coupling, but also induces supplementary elastic forces. In this example, the actuator is placed at a distance L from the pivot, and the load at a distance $3L$. This way, the displacement of the load is three times higher than at the actuator side, while the output force is amplified by the same factor of 3.

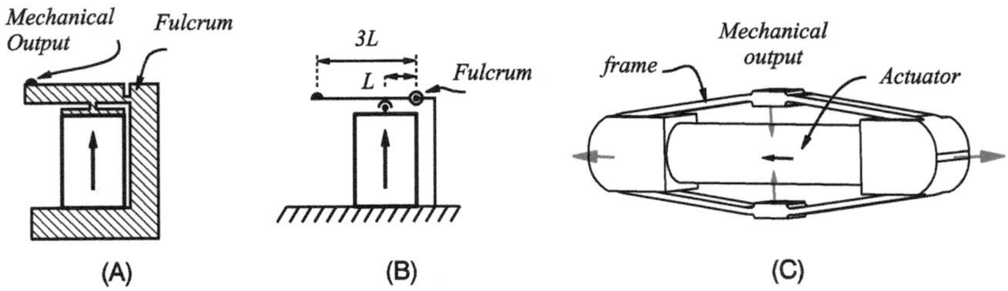

Figure 1.8 Amplified actuator: (A) example of a realization, (B) kinematic scheme, and (C) implementation with an elastic frame.

Table 1.4 Characteristics of some amplified actuators. Source www.noliac.com.

Designation	$W \times L \times H$	Max deflection	Max force	Stiffness
NAC2641	$14 \times 70.5 \times 26.1$ mm^3	300 μm	250 N	1.3 N·μm^{-1}
NAC2643	$14 \times 102.4 \times 28$ mm^3	625 μm	250 N	0.9 N·μm^{-1}
NAC2645	$14 \times 134.2 \times 30.6$ mm^3	950 μm	250 N	0.7 N·μm^{-1}

In some other realizations, the piezoelectric actuator is embedded into an elastic frame made of metal, with two arms along the axis of the actuator (Fig. 1.8 (C)). When the voltage is applied, the latter expands along its axis. Due to the particular form of the frame, the two arms stretch out, forcing the mechanical output to move along an axis perpendicular to the actuator's axis.

Hence, as with benders, amplified actuators can produce larger stroke with respect to the piezoelectric actuator they use, but at the expense of a stiffness reduction and of their force capacity. The characteristics of some devices are presented in Table 1.4.

1.1.3 Resonating actuators

Mechanically assisted ultrasonic[3] vibrations play an important role in processes. Dentists use ultrasonic oscillations of a tip to remove tartar from the teeth with their *ultrasonic* scaler. In food industry, mechanical knives vibrate at a frequency around 30 kHz to easily cut pieces of a cake without squashing them. And portable nebulizers use an ultrasonic wave to shake drops of liquid and create mist.

[3] Ultrasonic vibrations occur at a frequency greater than 20 kHz, which is the human limit of hearing.

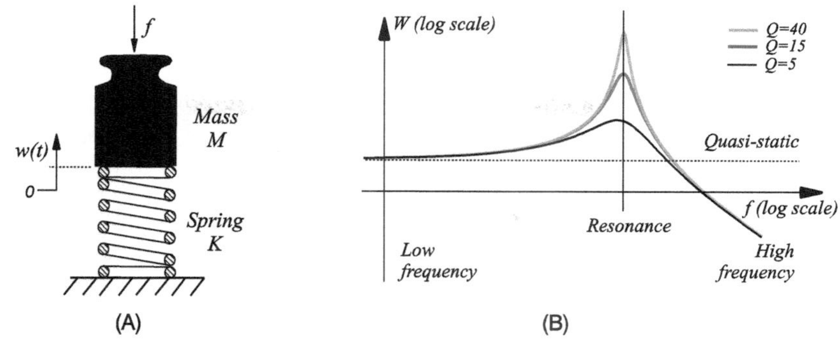

Figure 1.9 Mass–spring system: (A) a typical system, (B) frequency response to forced vibrations.

Because these applications use oscillations, the actuators can use the effect of *resonance*. Resonance occurs when the driving forces of a vibrating system produce a greater oscillation amplitude at a preferential vibration frequency than at an other one. And the aforementioned processes use harmonic oscillators put in resonance by a piezoelectric actuator. To obtain this, one solution consists in creating a resonance between a spring and mass, and to assist the vibrations with a piezoelectric actuator, the mass and stiffness of which can actually participate in the resonance.

If we consider a mass M mounted on a spring of stiffness K, we create a harmonic oscillator as depicted by Fig. 1.9. Its resonance angular frequency is equal to $\omega_0 = \sqrt{K/M}$. If the system is moved away from its equilibrium point, natural oscillations begin, and slowly decrease in amplitude due to the damping in the system. Vibrations can, however, be forced by external forces, denoted by $f(t)$, at a pulsation ω and amplitude F, i.e., $f(t) = F\sin(\omega t)$. Then, the mass M, with its position denoted by $w(t)$, is oscillating at the same pulsation ω, with an amplitude W. The vibration amplitude not only depends on the amplitude of the force F, but also on the vibration frequency ω. This is shown in Fig. 1.9 (B). At low pulsation, the elastic behavior of the spring is predominant, and the vibration amplitude doesn't depend a lot on pulsation. For the quasistatic operating mode, where the pulsation is almost equal to 0, the vibration amplitude is proportional to F. In contrast, at high frequency, the inertia of the mass is predominant: the vibration amplitude decreases with pulsation. In-between the resonance occurs, for a pulsation where the elastic behavior compensates the inertia of the mass.

At the pulsation of resonance, the vibration amplitude is limited only by the damping of the system, and can be much higher than the vibration obtained for quasistatic operation. The ratio between the amplitude at resonance and in quasistatic operation is named *quality factor* and is denoted by Q. The theory of oscillators also shows that the Q-factor is equal to the ratio between the resonant pulsation and bandwidth. Hence, a high quality factor is an advantage, and a drawback: an advantage because for a given quasistatic vibration amplitude, it results in high vibration amplitude at resonance; and a drawback because it is more selective: with an ultrasonic actuator at 20 kHz having a Q-factor of 200, the frequency should by set with a precision of less than 0.5% to get the maximum performance.

In most cases the acoustic phenomena cannot be neglected, and wave propagation inside the mass–spring system occurs, leading to nonuniform displacement of the particles of the system. This happens when the size of the vibrating mass is not small compared with the distance traveled by sound in a half-period. In a stainless steel rod, for example, sound speed is equal to 5790 m/s. At 100 Hz, this leads to a traveled distance of 290 m. But at 10 kHz, it is 29 cm. Hence, in a 29 cm long mass–spring system, the speed of the particles are not uniform in the matter. As a consequence, the mass cannot be considered as stiff, and the mass–spring model is not accurate.

This is what happens in a Langevin transducer. A Langevin[4] transducer is usually built up with two bulk piezoelectric rings placed between a front and back masses, generally of aluminum or steel and clamped by a nut, as shown Fig. 1.10 (A). When supplied with a sinusoidal voltage at the right frequency, the piezoelectric elements create an acoustic wave, which propagates along the actuators. The more distant the particles from the center, the higher their displacement. This is depicted in Fig. 1.10 (C) for two particles, named *P*1 and *P*2, chosen at two different positions in the actuator. Fig. 1.10 (B) (depicts the amplitude of vibration *W* according to the position in the actuator: the two opposite faces expand and contract in phase, so the displacements are opposite to each other, which is taken into account by the sign of *W*. In the middle, no vibration occurs: this is called *a vibration node.* The vibration node is the convenient place to attach the actuator to the ground because this point does not vibrate: the connection is not stressed by the oscillations.

[4] Paul Langevin was a French physicist who pioneered the development of piezoelectric transducers during the twentieth century.

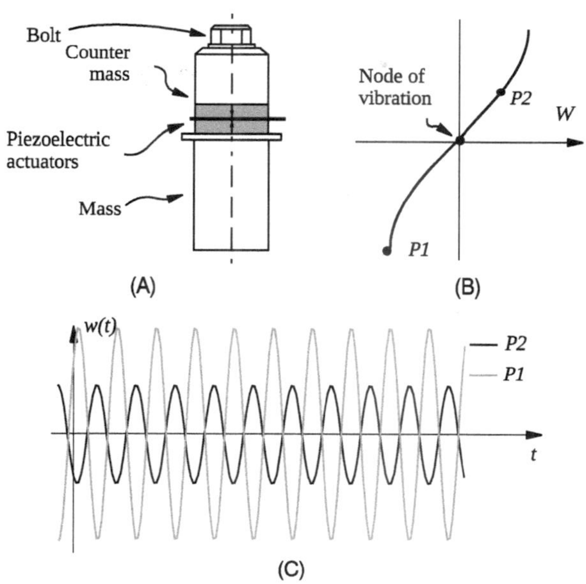

Figure 1.10 Langevin transducer: (A) principle, (B) vibration amplitude as a function of position, and (C) vibration as a function of time for two particles in the mass.

The resonant pulsation depends on the dimensions and material of the transducer, and they are calculated according to the requirement of the application. To precisely set the resonance frequency, the bolt is tightened. Tightening the bolt has the effect of increasing the resonance frequency. The same behavior is observed with a guitar string that is taut to get a higher tone.

Benders can also be used to create ultrasonic actuators. When supplied with a sinusoidal voltage, a bender deforms and bends, its tip producing an oscillation. If the frequency is increased, then the deformation of the bender changes, and several patterns appear at specific frequencies. Like with the Langevin transducer, each particle of the bender oscillates, but with a different vibration amplitude. The evolution of the vibration amplitude as a function of the position on the beam is called the *deformation mode shapes* of the bender, and Fig. 1.11 (A) presents the first deformation mode shapes of a thin bender, while Fig. 1.11 (B) presents the frequency response for the particle $P1$.

For the first modes and convenient design, the deformation of a bender can be high enough to be measured with a digital camera. However, to capture the deformation, it is necessary to synchronize the camera's trigger

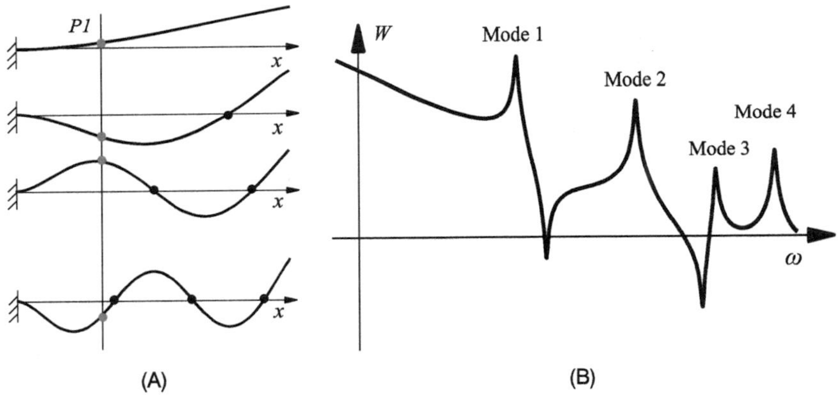

Figure 1.11 Deformation mode shapes of a bender: (A) the first four deformation mode shapes of a bender attached on one side, and free on the other side; • represents the vibration nodes, (B) frequency response of the vibration at position *P*1 (log scale).

on bender's motion, yet use a very small exposure time compared with the vibration's period. This can be very difficult to achieve when the resonance frequency is above 10 Hz.

To tackle this problem, we can use a stroboscopic setup presented Fig. 1.12. The principle is to use LEDs that are switched on at certain times synchronized on vibration. ON time is chosen to be small compared with the period of the vibration. Hence, the LED flashes the bender at instants which are the same in the vibration's period. By modifying the phase between the time to switch on and the vibration, it is possible to "freeze" the bender in a particular position of the vibration. A long exposure time can be applied then to capture this movement.

Fig. 1.13 presents a piezoelectric bender (V21B from Mide, www.mide.com) at its first bending mode, which occurs at 116 Hz. The two pictures are obtained for two phase values. For the top figure, the LED's flash occurs when the bender is in its neutral position ($w=0$), while for the bottom one, the bender is in it downward position, and we can see the difference of the bender's shape.

1.1.4 Piezoelectric motors

Piezoelectric and ultrasonic actuators are characterized by their capability to produce high forces and generate small displacements. These features helped to solve technological bottlenecks in applications needing lightweight and affordable actuators, because they can drive the mechani-

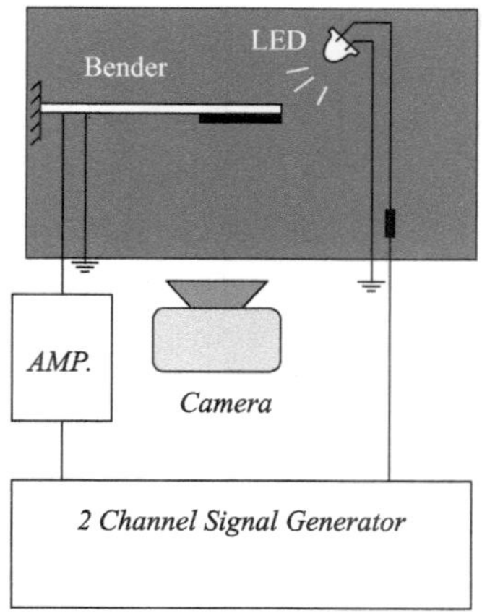

Figure 1.12 The stroboscopic setup; a 2-channel signal generator is used to adjust the phase between the bender's excitation and the LED flash.

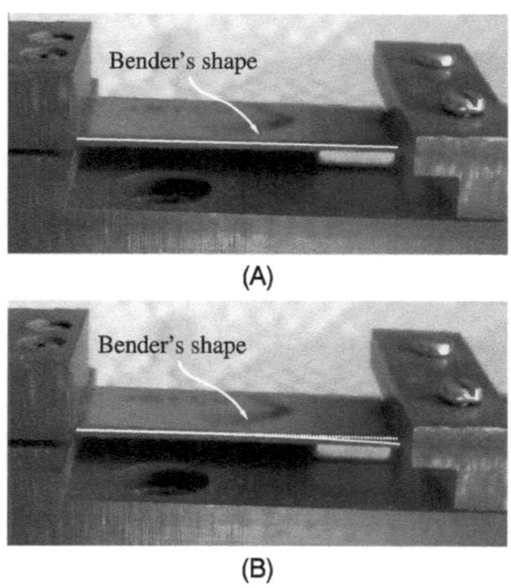

Figure 1.13 Picture of a bender with the stroboscopic setup: (A) phase reference, (B) phase shifted by 180°.

cal load directly, without gears, for example. The piezoelectric motors take advantage of these characteristics, but they are not limited in displacement: they can have a rotational or a translation output, yet produce high forces in a small volume.

In this section, we present two types of motor:

- PAD motors exploit the deformation of benders to achieve high resolution displacement of a rotor,

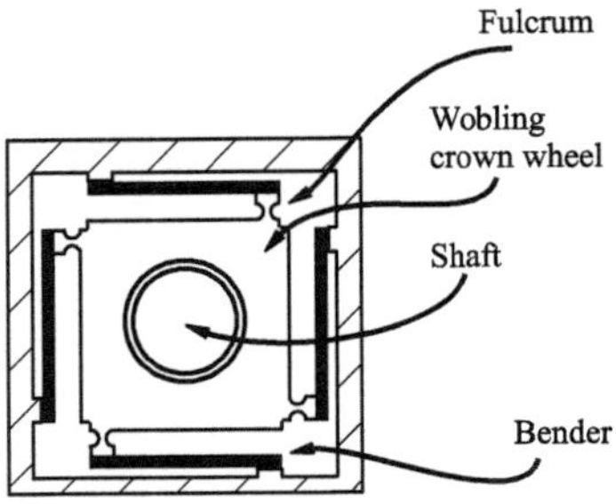

Rated speed	16.8 rpm
Rated torque	8.6 mNm
Driving Voltage	100 V
Driving Frequency	50 Hz
Weight	45 g

Figure 1.14 The PAD motor and its specifications, from www.noliac.com.

- Ultrasonic motors use oscillations of a stator to propel, by friction, a rotor pressed on it.

Some examples of such piezoelectric motors are given in the two followings sections.

1.1.4.1 PAD motor

PAD motors are low-speed, high precision motors. By principle, a wobbling crown wheel undergoes a circular translation. This movement is then converted into a rotational movement when the wheel is brought into contact with a geared shaft. Because the radii of the inner hole of the wheel and shaft are almost equal, a high speed reduction ratio is obtained between the circular translation and rotation of the shaft. The specifications are given in Fig. 1.14.

To obtain the wobbling movement, 2 sets of piezoelectric benders are used and driven by 2 sinusoidal voltages shifted by 90°, as shown in Fig. 1.15. However, in this application, the deformation of the benders is limited by the gap between the shaft and crown wheel. Therefore, once the crown wheel is in contact with the shaft, the voltage supplied to the bender increases the contact force, and not the displacement.

1.1.4.2 Ultrasonic motors

The Elliptec motor (Hemsel et al., 2006, Fig. 1.16) exploits an elliptic motion of a beam's tip, to push a slider (linear motion) or a wheel (rotation).

The elliptic motion is obtained by using two principles. First, a beam is excited at its first bending mode, and can oscillate up and down. Second, a linear actuator expands and contracts, thus producing a longitudinal movement. These two movements are in fact produced by the same actuator, thus they are synchronized, and the resonant frequency of the bending

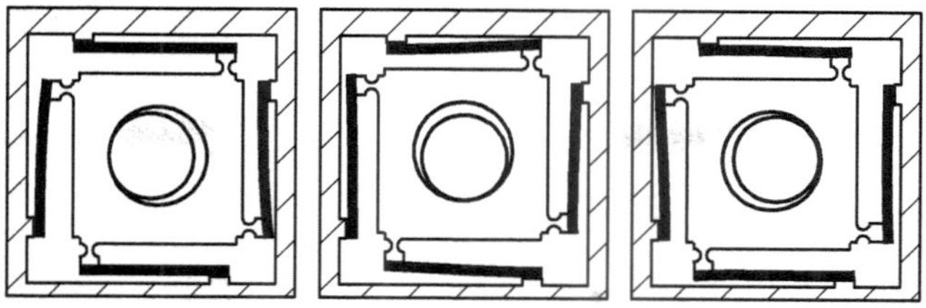

Figure 1.15 Typical movement of the crown wheel; the gap between the crown wheel and shaft has been enlarged for illustration.

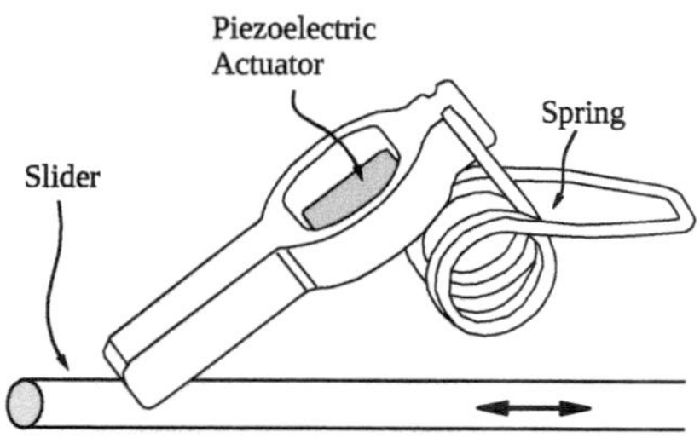

No-Load speed	350 mm·s^{-1}
Max. Blocking force	300 mN
Driving Voltage	7 V
Driving Frequency	81 kHz
Length	20 mm
Width	7.8 mm
Weight	1.2 g

Figure 1.16 Elliptec motor and the characteristic values of the X15G (from www.thorlabs.com).

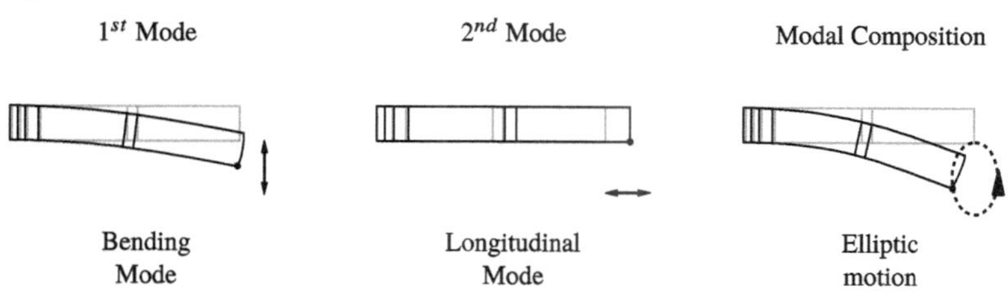

Figure 1.17 Elliptic motion of the motor's tip.

mode and the horizontal movement are matched, by adequately choosing the dimension of the actuator. The elliptic movement of the beam's tip results from the superimposition of the movements along the two axis, as depicted Fig. 1.17. Once in contact with the load, the beam's tip drives the latter. Then, the contact is broken and the tip flies back to its initial position, where it comes again in contact for a new cycle. Hence, the rotor and stator are in contact, but this contact is intermittent.

An other implementation of the elliptic motion is proposed in the *traveling wave ultrasonic motors (TWUM)*. A traveling wave is a vibration mode which nodes are moving at constant speed. The particles of the resonator

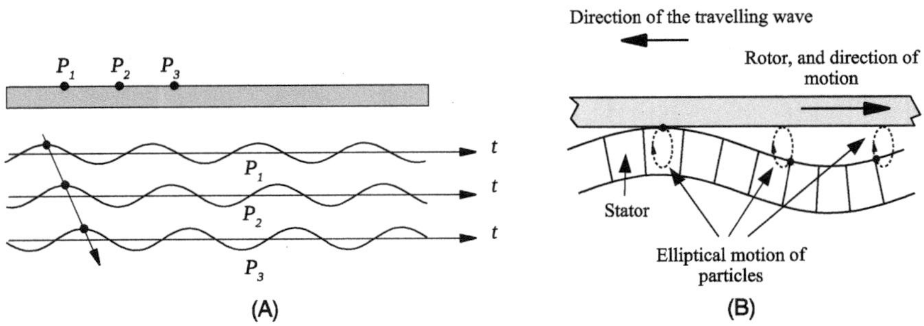

Figure 1.18 Definition of a traveling wave: (A) vibration amplitude for 3 positions, (B) elliptic motion of particles.

vibrate with the same amplitude, but with a phase shift which depends on their position, as depicted in Fig. 1.18 (A). When a traveling wave propagates in a beam, the vibration produces a normal and tangential displacement, thus the particles undergo an elliptic motion as shown by Fig. 1.18.

For these motors, a traveling wave is generated in a ring shaped stator, and seems to rotate at the stator's surface. Two vibration modes are composed to create the traveling wave; this modal composition will be studied in Chap. 3, but the important thing to notice is that each particle of the stator undergoes an elliptic motion. Hence, the particles at the top of the stator can propel the rotor firmly pressed on it, in the same way as already presented for the Elliptec motor; however, in the case of TWUM, the contact is permanent.

The key advantage of *TWUMs* is their high torque – low speed characteristics which allows *direct-drive* utilization: the motor's load is directly attached to the shaft, and no speed-reducer or gears are needed. Fig. 1.19 (A) compares the output mechanical torque as a function of weight for several typical electromagnetic motors (brushed and brushless DC motors) and for TWUMs. We can see that for the same rated torque ultrasonic motors are lighter: in direct-drive applications, TWUM gives more torque than an EM one, the ratio can be 10 for the same weight. However, when we compare the mechanical power as a function of weight, TWUMs are heavier than EMs. Therefore, TWUMs are usually not competitive in application with high power demand.

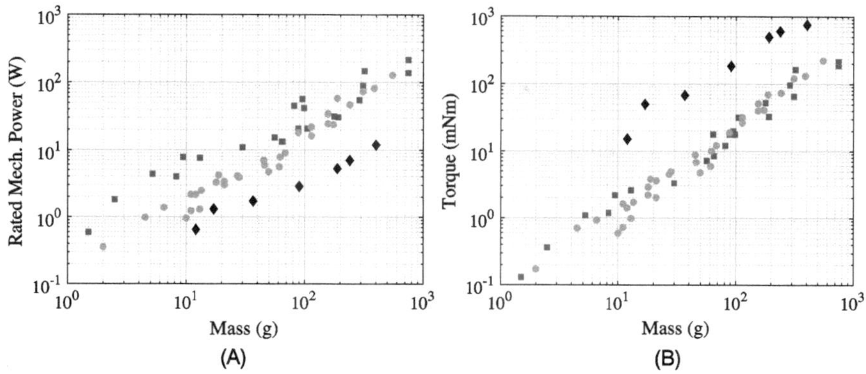

Figure 1.19 Comparison of TWUMs with typical electromagnetic motors: ♦ TWUM, ■ EM brushless motor, • EM DC motor; (A) Rated torque as a function of mass and (B) rated mechanical power as a function of mass.

1.2. Applications

The objective of this section is to give examples of applications using piezoelectric actuators.

1.2.1 Precise positioning

In industry, precise positioning refers to an average accuracy better than 10 μm (Sato, 2006). Such accuracy is fundamental in industrial machine tools, measurement systems and semiconductor manufacturing systems.

In atomic force microscope (AFM), for example, the deflection of a laser beam by a cantilever is measured. The cantilever is equipped with a sharp tip which scans a specimen surface; in another type of microscope, the surface tunneling microscope (STM), the current flowing in the tip is measured. To scan the specimen, one solution consists in using a tube made of piezoelectric material, with sectorized electrodes as presented in Fig. 1.20. When we supply a voltage on the *X* electrodes (up to 1 kV), the piezoelectric tube bends in the *X* direction, producing a displacement of the sample. When supplying voltage on the *Y* electrodes, a displacement in the *Y* direction occurs.

To precisely place the sample under the microscope's tip, a piezoelectric motor can be used. The Pan method, from its inventor's name (Pan et al., 1999), consists in using piezoelectric legs, built up with piezoelectric stacks working in shear mode to move a sapphire prism integral

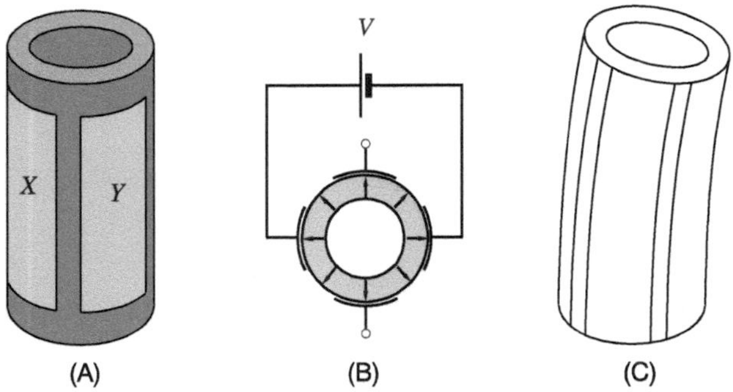

Figure 1.20 Tubular XY scanner for AFM microscope: (A) PT230, producing up to ±30 μm (photo: PI.com), (B) voltage connection for a deflection in the *X* axis, and (C) deformed tube.

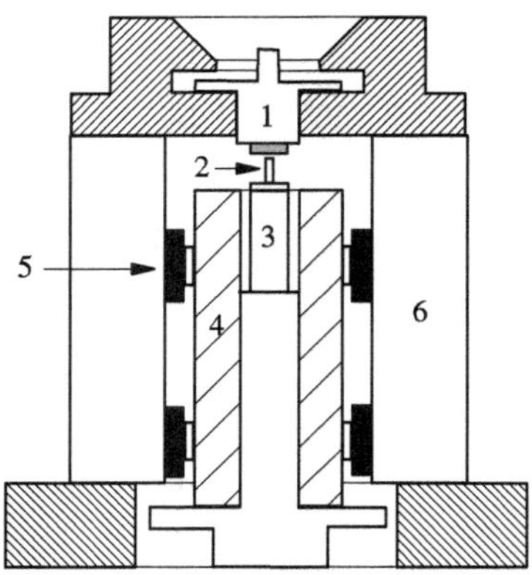

Figure 1.21 A linear piezoelectric motor to position the sample in an STM (from Pan et al., 1999). A cross-section with: 1, sample holder; 2, tip; 3, XY scanner; 4, sapphire prism; 5, shear piezostacks; and 6, Maccor body.

to the tip's holder (Fig. 1.21). To obtain a coarse motion, a voltage is applied successively on each leg. This way, each leg shears without moving the prism, because it is blocked by the other legs. When all the legs have been deformed, the voltage simultaneously and smoothly returns to 0: all the legs accompany the prism altogether. Hence, the tip is moved step by step. One advantage of the solution is to block the tip without electrical power, and to allow coarse displacement with a high resolution.

Optical applications are also targeted by piezoelectric actuators. They are often used to deflect a laser beam precisely.

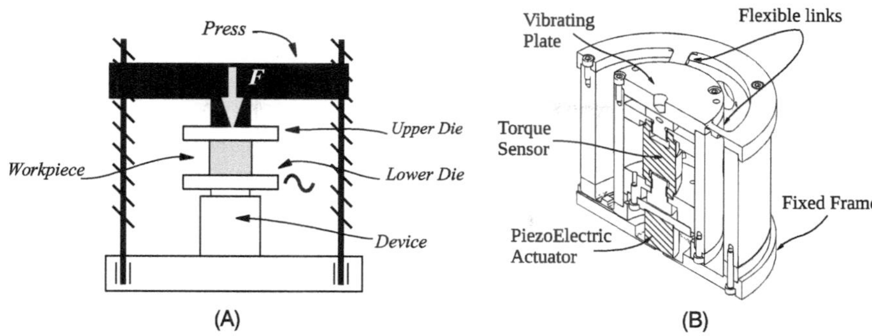

Figure 1.22 Assisted forging mechanism: (A) principle and (B) the device, able to produce vibration at 30 Hz with an amplitude of 20 μm, under a forging load of 10 kN reducing by 500 N the pressing force *F*.

1.2.2 Industrial processes

Vibration is often used to improve industrial processes. High frequency vibrations are helpful in industrial cleaning, because the vibration propagates in a liquid, producing its cavitation which removes dirt and contaminants from the surface of an object (Tangsopa and Thongsri, 2019). Also, vibration is used to compact powder: the high-frequency shaking of the powder evacuates the air bubbles between the grains, ensuring compaction.

We can also cite the assisted forging process, which consists in using a vibrating plate, and pressing on it with a workpiece to give the latter a specific shape. Vibrations of a few micrometers are enough to have a positive effect in terms of forging load and shape. But the forces needed are high, and the actuator has to produce forces as high as 10 kN. In such applications, the assembly is critical, because parasitic forces (forces resulting in shear stresses inside the actuator) have to be reduced; Fig. 1.22 gives an example of a typical implementation (Khan et al., 2016).

1.2.3 Haptic stimulators

Haptic stimulators are actuators which are able to convey information to a human through the sense of touch. One example is the vibration produced by a handheld smartphone to alert the arrival of a call. But more sophisticated haptic stimulators exist. In a *High Fidelity Haptic Stimulator*, a vibration is produced by a piezoelectric bender, on which a seismic mass is bonded. When the actuator is supplied with a sinusoidal voltage, the mass vibrates, shaking the whole body of the device (Fig. 1.23). Compared with electromagnetic actuators, which can achieve the same task, haptic stimulators

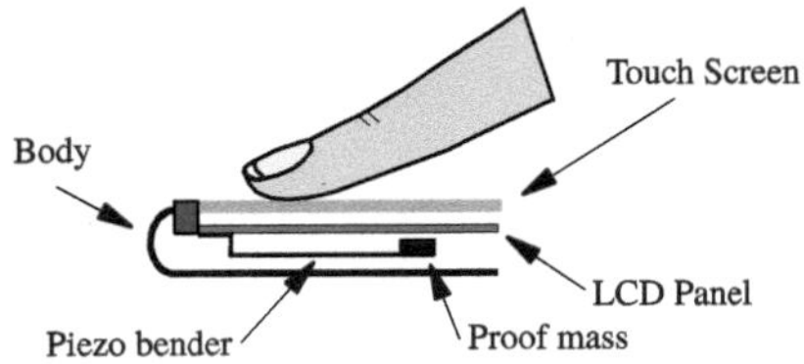

Figure 1.23 Principle of piezoelectric haptic actuators.

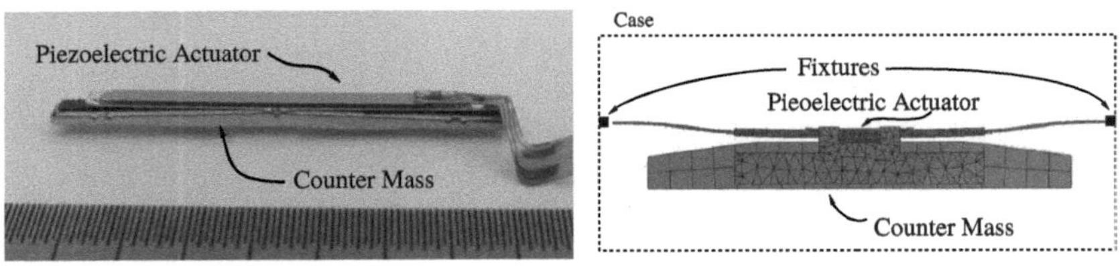

(A) (B)

Figure 1.24 Principle of PHA379060: (A) dismantled view without the case and the fixtures, (B) deformation of the actuator.

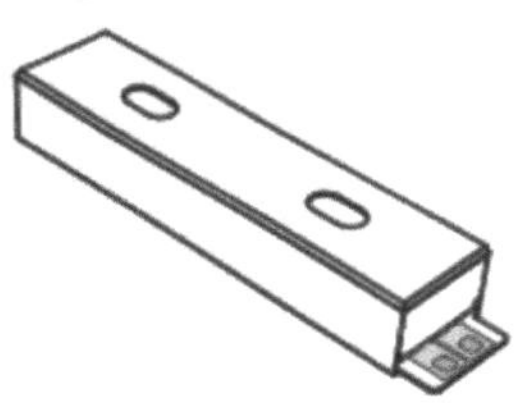

size	$6 \times 9 \times 37$ mm^3
Voltage	20 Vpp
Bandwidth	165 Hz
Power	220 mW

Figure 1.25 Typical performances of the piezoelectric haptic actuator PHA379060 (www.mpluskr.com).

offer a higher bandwidth, but require higher voltage and consume more power.

The PHA379060 from *mplus* (Republic of Korea) is an example of a piezoelectric haptic stimulator, which is built up with a thin multilayer piezoelectric actuator, bonded onto a substrate of steel, as presented by Fig. 1.24 (A). It is attached on its two edges to the case; consequently, a symmetrical deformation is obtained as depicted in Fig. 1.24 (B) when a voltage is applied. This design is more robust than that of the bender. A typical specification for this device is given in Fig. 1.25.

In a *Surface Haptic Display*, the top surface of an LCD screen is covered by a transparent glass plate, equipped with piezoelectric monolayer actuators. When supplying a sinusoidal voltage to the ceramic, a vibration of $\simeq$1 µm is produced at 60 kHz, which corresponds to a resonance frequency.

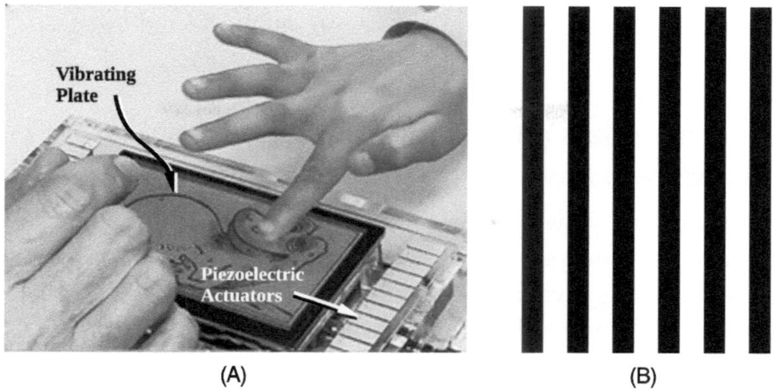

Figure 1.26 A prototype of a *Surface Haptic Display* (Vezzoli et al., 2016) (A) and a friction map to give the illusion of touching a grating surface (B).

This vibration is not felt by a user touching the vibrating plate: it is too high in frequency and too low in amplitude to be detected by a human finger. However, the fingerpulp is in intermittent contact with the vibrating plate, and this reduces the friction between the plate and the finger when it is moving: the user feels as if the plate is lubricated by a thin film of lubricant. If the voltage is stopped, then the friction reduction is removed, and the finger is blocked. When modulating the vibration amplitude as a function of fingertip's position, a large number of tactile effects can be programmed. Fig. 1.26 presents such a haptic display in use; the user can feel details of the frog's skin as if it was real on the screen. For that purpose, a friction map is created, such as the one depicted in Fig. 1.26 (B): friction is decreased (white zones) or not (black ones) as a function of finger's position to make details appear on the screen.

1.3. Variability of the operating point of piezoelectric actuators

The previous sections in this chapter have shown how piezoelectric actuators can indeed convert electrical energy into mechanical energy. However, it is also essential for design engineers to know their advantages and practical issues in order to integrate them inside the systems they design. Compactness, force or torque-to-mass ratio, and form-factor are among their most appreciated qualities. But the actuators also have to be reliable, and the operating point should not be too much affected by external factors.

1.3.1 Effect of material's nonlinearity

1.3.1.1 Increased power losses

The previous section has shown the principles of the most common piezoelectric actuators. A basic piece of piezoelectric material works like a hydraulic cylinder: supply a voltage to it, and it will move a load a certain distance. The relation between this distance and the voltage can be considered almost proportional for rough considerations. However, the material's properties of PZT are not linear, nor is the aforementioned relationship, with hysteresis inducing a memory effect. Control strategies have been proposed to compensate for this nonlinearity, and will be presented in Sect. 1.4.

The nonlinearities can be classified relative to:

- elastic behavior,
- dielectric behavior, and
- piezoelectric behavior.

They lead to increased power losses. To illustrate this, we consider a pure elastic material following a deformation cycle. The energy lost in one cycle is defined as

$$E_e = \int_{\text{cycle}} f \, \mathrm{d}w, \tag{1.1}$$

where f is the external force, and w is the actuator's stroke. If we draw (Fig. 1.27) w as a function of f, we obtain a linear curve for a pure elastic material; Eq. (1.1) yields 0. But for a nonlinear material, hysteresis appears, and (1.1) is proportional to the surface inside the hysteresis. Hence, at each deformation cycle, some energy is lost inside the material, and dissipated into heat. If the cycle is repeated periodically, the heat may be not fully dissipated and the temperature increases. In some modeling, this process is described by an internal damping in addition to the material stiffness.

Piezoelectric and dielectric behaviors are also nonlinear, with the same consequences: energy loss at each cycle and temperature rise of the piezoelectric material. For the latter, and for a pure sinusoidal voltage v at a pulsation ω such that $v(t) = V\sin(\omega t)$, the dielectric power losses P_d are linked to the reactive power in the blocked capacitance C by (see Table 1.2, page 6)

$$P_d = \frac{1}{2} C \omega V^2 \tan(\delta), \tag{1.2}$$

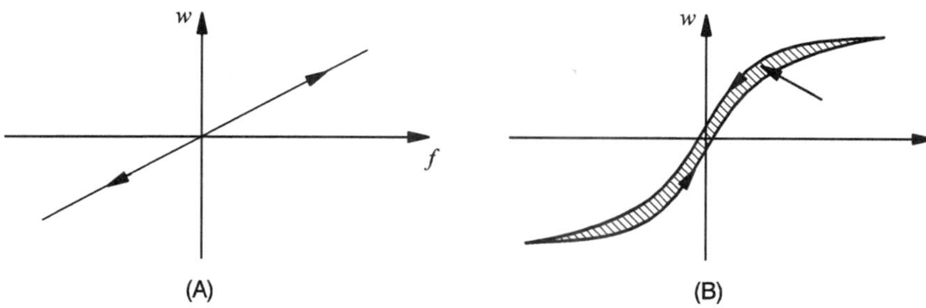

Figure 1.27 Stroke w as a function of force f applied to the material: (A) linear material, (B) nonlinear material; the energy loss E_e is proportional to the area inside the hysteresis.

where $\tan(\delta)$ is a parameter given by the manufacturer of the piezoelectric material. From an electrical point of view, the dielectric power losses can be modeled by a resistor R as shown Fig. 1.3, page 5, and Eq. (1.2) yields

$$R = \frac{1}{C\omega \tan(\delta)}. \tag{1.3}$$

This is why, due to temperature limitation, the frequency of the cycle is limited, because if ω increases, then R decreases, and thus the power dissipated in the material increases for a given voltage. Therefore, high frequency operation should lead to voltage reduction.

1.3.1.2 Saturation effect in an ultrasonic actuator

The nonlinearity of the material also has particular consequences on an ultrasonic resonator. To illustrate the effect of the nonlinearities of the material on the dielectric losses, we consider in Fig. 1.28 an ultrasonic actuator built up with a $83 \times 49 \times 2$ mm^3 copper plate, onto which a thin layer of piezoelectric material is bonded. This plate shows a bending mode at 31 kHz. The nonlinearities of the material result in a saturation of the vibratory amplitude: compared with a linear material, the vibration amplitude is lower, and the difference increases as the voltage increases. The dielectric losses show a v^2-dependence, as demonstrated by Eq. (1.2). If we consider an operating point at 0.75 µm, the experimental sample requires a voltage $v = 15$ V, leading to 3 W of power losses. If the material were linear, without the saturation effect, the same vibration amplitude would require $v = 11$ V, leading only to 1.5 W of power losses. The difference between the linear modeling and the actual material dramatically increases as the amount of required power increases, and for the actual material, high vibration amplitude leads to a very high amount of power losses.

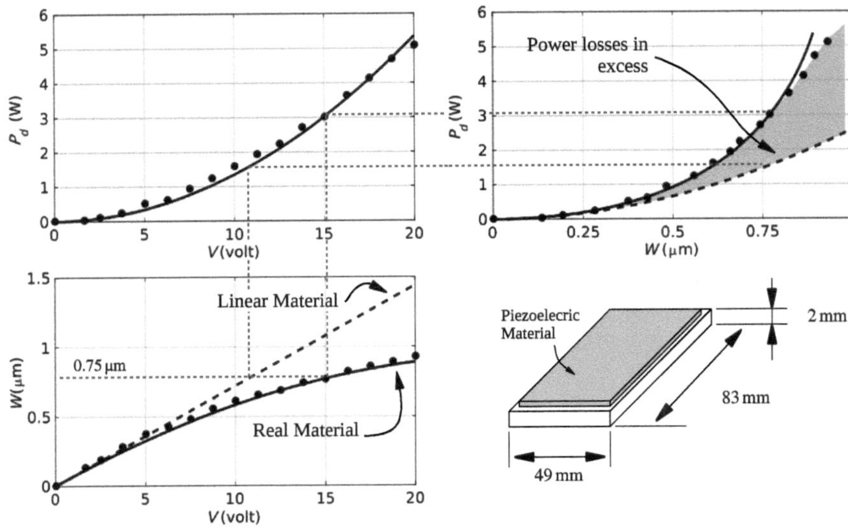

Figure 1.28 Relation between vibration amplitude W, voltage V and dielectric power losses P_d in a nonlinear material: – modeling, • experimental data, - - linear modeling.

Hence, it is very important to use the piezoelectric resonator in its linear region, where the relation between the voltage and vibration amplitude is linear. When the behavior is really nonlinear, leading to a saturation of the vibration amplitude, the dielectric power losses are in excess, and the additional voltage level which is added in an attempt to increase the vibratory amplitude serves only to increase the dielectric losses. In this region, the resonator is more a heat generator, than a resonator (Uchino, 1997). The design of the resonator is then crucial to ensure that for the rated vibration amplitude, the resonator is still in its linear region. This is achieved by adequately chosen size, thickness and material.

1.3.1.3 Resonance frequency shift

The nonlinear behavior of the material also has an effect on the frequency response of a piezoelectric resonator. In general, we can consider that the equivalent stiffness K of the resonator and the damping coefficient Q are affected by the vibration amplitude: they both decrease as the vibration amplitude W increases, as depicted in Fig. 1.29.

To understand the nonlinear effect produced by these variations, we draw in Fig. 1.30 the frequency response of the resonator **at constant vibration amplitude** W_1. To achieve this practically, we adjust the level

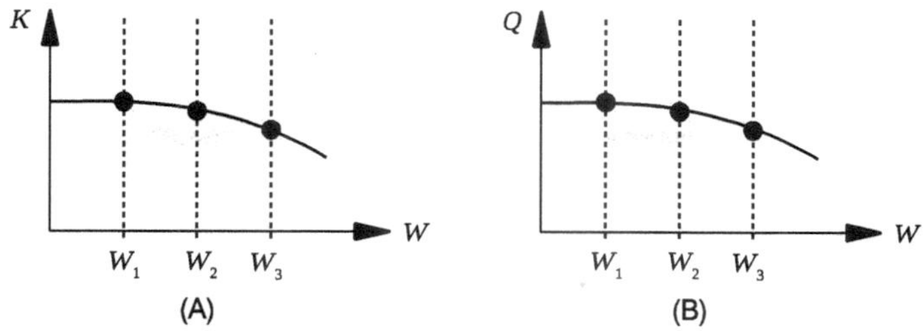

Figure 1.29 Variation of K and Q for an ultrasonic resonator.

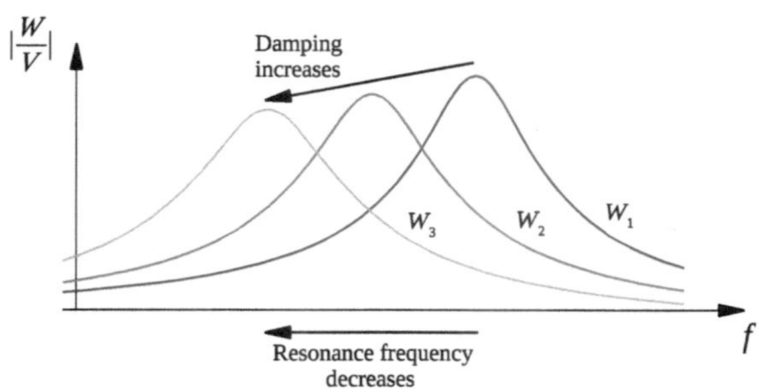

Figure 1.30 Frequency response at a constant vibration amplitude; $W_1 < W_2 < W_3$.

of voltage V in order to obtain $W = W_1$ for each value of the frequency, and we report the value of the ratio $\|\frac{W_1}{V}\|$. We repeat this for two other vibration amplitudes, W_2 and W_3, with $W_1 < W_2 < W_3$. We can observe in Fig. 1.30 that the nonlinearities result in:

- a frequency shift of the resonance frequency toward the low frequency when W increases,
- an increased damping of the system.

Practically, piezoelectric resonators are often characterized with their frequency response obtained *at a constant voltage* V. However, the resulting curve strongly differs from the frequency responses at a constant vibration amplitude. To understand why, we start from the curves in Fig. 1.30 and we multiply them by V. On these new curves, we note the values of the vibration amplitude which has been set. The resulting curve W at constant voltage V is obtained by connecting all the points between them, as depicted in Fig. 1.31.

The frequency response *at constant voltage* differs from the typical frequency response of a linear resonator: the top of the curve seems leaning. In general, for a given frequency f, there is only one solution to W. But for $f_1 < f < f_2$, the curve at constant voltage shows 3 solutions to w. This results

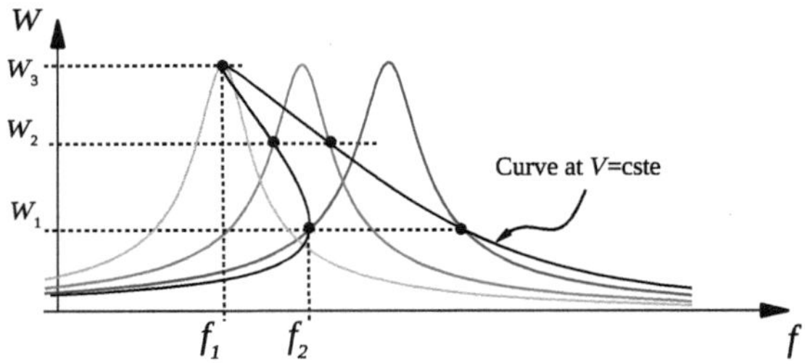

Figure 1.31 Frequency response at constant voltage amplitude.

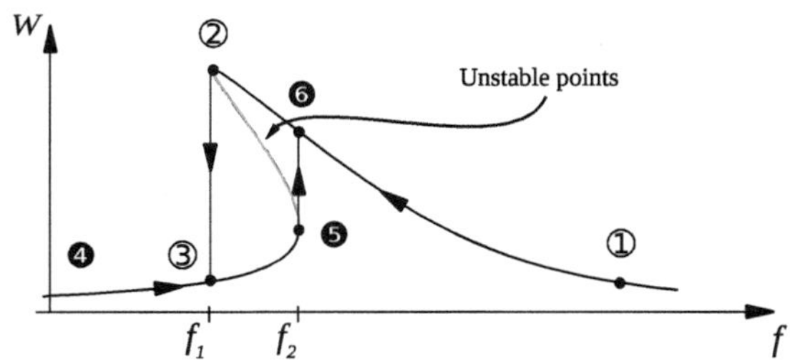

Figure 1.32 The pull-out phenomenon.

in a nonlinear behavior with a hysteresis effect, as presented in Fig. 1.32, and named the *pull-out* phenomenon.

If we start from a high frequency, denoted by point ① in Fig. 1.32, and if we decrease f, then the vibration amplitude increases, and can reach its maximum at point ②. If we decrease f even more, then the vibration amplitude drops abruptly to a much lower value at point ③: the curve $W(f)$ is discontinuous at $f = f_1$.

If we now increase f from the point ❹, the vibration amplitude increases and, at point ❺, the vibration amplitude increases abruptly to a higher value at point ❻. This high value is not the maximum, however, which is obtained at point ③ only.

Hence, the consequences of the nonlinearity of the material on the frequency behavior of a piezoelectric resonator working at a constant vibration amplitude are twofold:

- a hysteresis appears, and the vibration amplitude depends not only on f but also on history,
- the maximum vibration amplitude is obtained when decreasing the frequency and not increasing it.

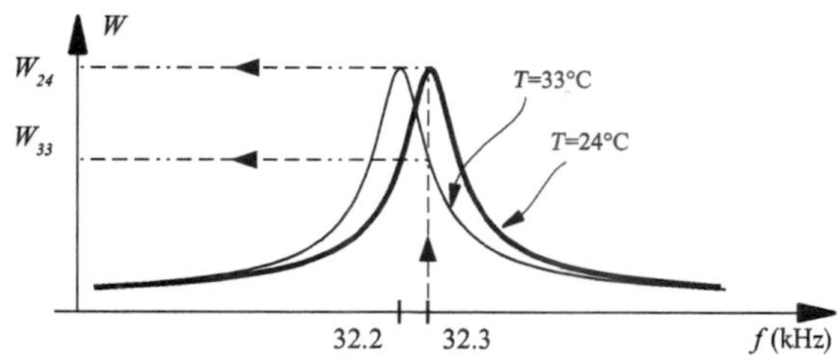

Figure 1.33 Effect of the temperature on the vibration amplitude at constant frequency and voltage.

1.3.2 Effect of the temperature

Temperature plays an important role in piezoelectric actuators. First of all, temperature increase often leads to thermal expansion of the actuator itself, which is of the same order as that produced by the piezoelectric effect. Hence, thermal expansion can cancel the elongation of a bender, amplified actuator or piezoelectric stack, which are therefore very sensitive to the temperature. A typical thermal expansion coefficient can reach 3 ppm/K. The INVAR alloy is thus often used in conjunction with piezoelectric material, because this alloy (made with iron and nickel) exhibits almost no thermal expansion. This is not enough, however, to maintain the precision of the positioning, and in some cases, a closed-loop control of the position is achieved.

Thermal expansion is not the main problem with ultrasonic actuators. Indeed, in that case, the deformation of the actuators oscillates around a static deformation, which variation with temperature – which is very slow – has no consequence on the process they are used for. However, temperature softens the materials, and the resonator's stiffness decreases when the temperature increases. This effect leads to the decrease of the actuator's resonance frequency. The variation is relatively small (around 300 ppm/K). But the consequence can be dramatic if the resonator has low damping. To illustrate this point, Fig. 1.33 shows the simulated frequency response of a Langevin transducer at two temperatures: at 24°C, the resonance is at 32.3 kHz, and at 33°C the resonance is at 32.2 kHz. Therefore, supplying a voltage at a constant frequency of 32.3 kHz would lead to a high vibration amplitude at 24°C because it corresponds to the resonance, but a very low vibration amplitude at 33°C because it falls outside the actuator's bandwidth. Hence, it is often necessary to add a resonance frequency tracking system to the piezoelectric resonator in order to automatically adapt the voltage frequency to the operating conditions.

1.3.3 Effect of the mechanical load

The deformation of a piezoelectric actuator not only depends on the voltage supplied to it, but also on the mechanical load it drives. We look at four specific cases to illustrate this point:

- When the actuator has no load connected to it, it can deform freely, and the nominal deformation occurs for the nominal voltage. This is called the *no-load* operation.
- When the actuator is blocked by a frame with infinite stiffness – or a stiffness much higher than the actuator's stiffness – it cannot deform: increasing the voltage has no effect on the deformation. This is called the *blocked* operation.
- When a constant force is applied on the actuator, the deformation at a given voltage v differs from the no-load operation. This loss of travel is due to the actuator stiffness.
- When pure inertia is attached to the actuator, resonance occurs as already described in Sect. 1.1.3.

A generally admitted model of the actuator's elongation w as a function of voltage v and force applied to load f is given by

$$w = \alpha v - \beta f, \tag{1.4}$$

where α and β are parameters, which depend on actuator's geometry and material. They are constant in the ideal case (no drift, no hysteresis, linear behavior). This model is convenient to describe the actuator's behavior under the aforementioned cases.

Uncer no-load operation, the force f equals 0, and thus the relation between the voltage and stroke is $w = \alpha v$. Therefore, the characteristic $w(v)$ is a straight line with slope α.

Under blocked operation, the stroke w equals 0. The actuator pushes the frame with a force f which is given by $f = \frac{\alpha}{\beta} v$.

Under constant force, Eq. (1.4) shows that the characteristic $w(v)$ is a straight line with slope α, the same as for the no-load operation. However, the stroke is reduced by a constant quantity βf, which is the loss of travel.

When the load is a spring, the force f is given by $f = Kw$ where K is the spring's stiffness. Therefore, stroke w is now given by $w = \frac{\alpha}{1+\beta K} v$. Yet, the characteristic is a straight line, with the reduced slope compared with the no-load operation. Here, the loss of travel is not constant.

For the inertia case, when a mass M is actuated by the piezoelectric stack, Newton's law gives $M\ddot{w} = f$, and Eq. (1.4) leads to $\beta M\ddot{w} + w = \alpha v$.

Table 1.5 The stroke of a stack actuator loaded by different types of load.

Type of load	**Example**	$v(t)$ **and** $w(t)$	**Characteristics**
no load	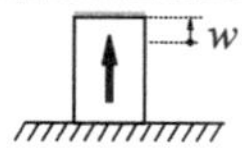	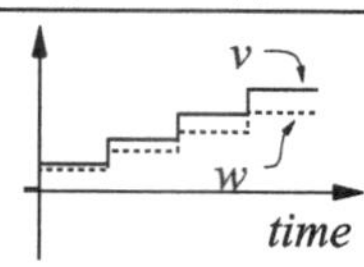	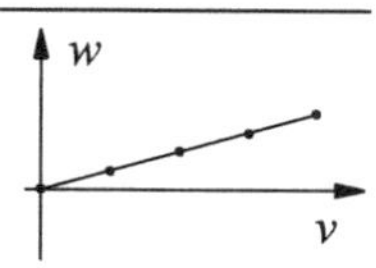
force F	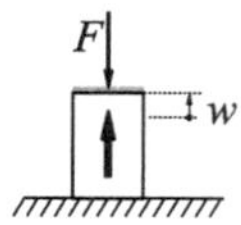	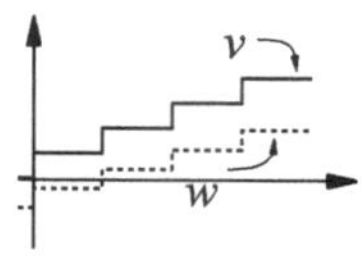	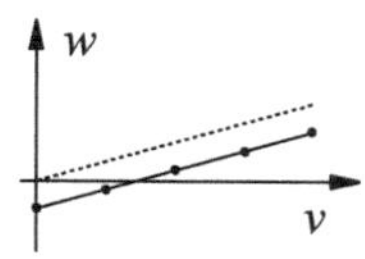
spring	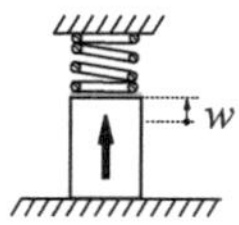	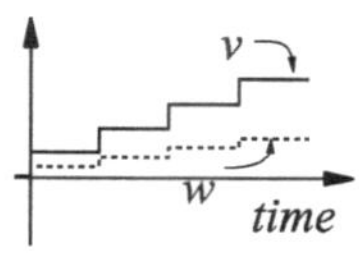	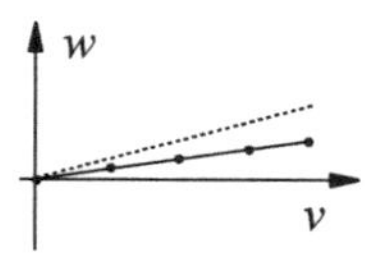
inertia	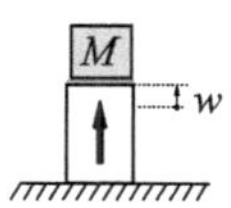 (Gravity neglected)	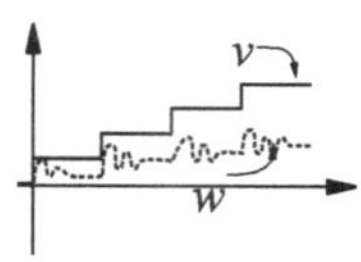	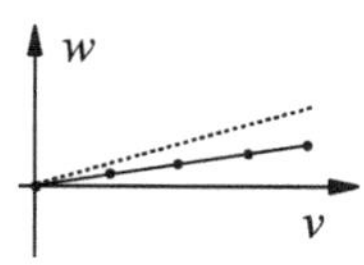

The steady state ($\ddot{w} = 0$) results in the same operating point as the no-load operation, but the second-order term can lead to oscillation in response to step variation of the voltage.

Table 1.5 presents some of the aforementioned cases. We consider a multistack actuator with different types of load, supplied with a step varying voltage. We present the resulting deformation as a function of time and in the voltage–displacement plane. This figure illustrates the fact that the voltage does not necessarily control the displacement, in particular when the type of the load is not constant.

The effect of the load of an ultrasonic actuator is different. For example, we consider the Langevin transducer of Fig. 1.10 which is used to burn living tissues in medical applications as in (Liu et al., 2015). The mechanical load can be modeled as a stiffness which is added to the actuator's stiffness. The resulting effect is that the frequency at which the vibration amplitude is maximum is slightly decreased compared with the frequency of the actuator when it is alone. In other words, the resonant frequency of the system composed of the actuator and its load is different from the resonant frequency alone.

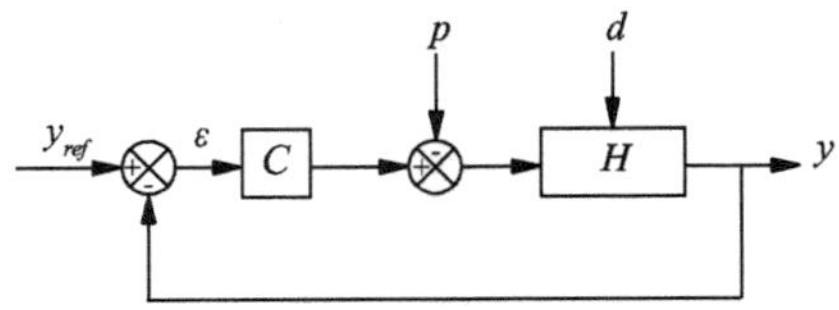

Figure 1.34 The scheme of closed-loop control.

1.4. Control of piezoelectric actuators

The previous section has shown that the behavior of the piezoelectric actuators can be affected by external factors like load and temperature, and internal factors like material's nonlinearity. In order to compensate their effects, the system's input (voltage v or current i) can be tuned. This can be made with a *controller*. The following section gives the basics of the control theory before giving examples of piezoelectric actuators and motors.

1.4.1 Control basics

The input command $u(t)$ of a system should vary according to a reference setpoint y_{ref} for its output $y(t)$ and all the disturbances $d(t)$. This can be achieved *by an open loop* provided that it is possible to have perfect knowledge of the system and its disturbances. Unfortunately, most of the time, this is difficult or impossible, and we only have an estimate of the system's behavior. This estimate is not enough to deduce by inversion the input u to apply.

This is why we use *closed-loop control* instead. In closed-loop control, u depends on the error $\varepsilon(t) = y_{ref} - y(t)$. A controller C is then designed in order to tune the input command u in order to obtain $\varepsilon = 0$ or $y(t) = y_{ref}$. The general structure of a control loop is given in Fig. 1.34, and for the controller C design we refer to the control theory; we will not go too deep into the details about this procedure in this book. Hereafter, p refers to the perturbation (load force, for example) which acts like an input, and d to the disturbances (temperature, for example) which can modify the system's behavior H.

The design of a controller follows several steps. First, it is essential to clearly define a list of design requirements which will have to be fulfilled in operation by the controller. For example, this can be the elimination of the *static error* (no error in steady state), a maximum *response time* (the time needed to obtain 95% of the reference setpoint), robustness besides external factors (the output is not affected by change in temperature, load, etc.). This step needs experience, and is guided by the application.

The second step consists in choosing the corrector C. There exist many correctors, each with its own performance. We can cite *PID* (Proportional, Integral, Differential) controllers, RST and H_∞ controllers. They all are linear controllers. Specific features can be added, like *self-tuning* or *auto-adaptation*. This will not be used in this book. Other controllers are nonlinear, like fuzzy controllers, neural networks.

In this book, we will use *PI* controllers, because they are widely used in industry, and easy to implement. The design of linear controllers relies on linear modeling of the system to control. Hence, the third step is to tune the controller's parameters according to the system's model in order to obtain the required performances defined in Step 1.

The general form of a PI controller is given using the *Laplace transform* (Wikipedia contributors, 2018):

$$C(s) = K_p + \frac{K_I}{s}, \tag{1.5}$$

where K_p and K_I denote the proportional and integral terms, respectively. We now give two examples of the analytical controller design.

The system is approximated by a pure gain: If the system is approximated by a pure gain such that $H(s) = G$, a pure integral corrector is used and $K_p = 0$. The *Open-Loop Transfer Function* $OLTF(s)$ is given by

$$OLTF(s) = C(s) \times G = G\frac{K_I}{s}. \tag{1.6}$$

The *Closed-Loop Transfer Function* $CLTF(s)$ is then calculated as

$$\begin{aligned} CLTF(s) &= \frac{y}{y_{ref}} = \frac{OLTF}{1+OLTF} = \frac{G\frac{K_I}{s}}{1+G\frac{K_I}{s}} \\ &= \frac{1}{1+\frac{s}{K_I G}}. \end{aligned} \tag{1.7}$$

Hence, the transfer function $\frac{y}{y_{ref}}$ is a first-order type equation, which means that in a closed loop, the system behaves like a first-order system. In particular, it is well known that the response time t_r of a first-order system is equal to $t_r = 3\tau_{BF}$, with τ_{BF} being named the *time constant* and equal to

$$\tau_{BF} = \frac{1}{K_I G}. \tag{1.8}$$

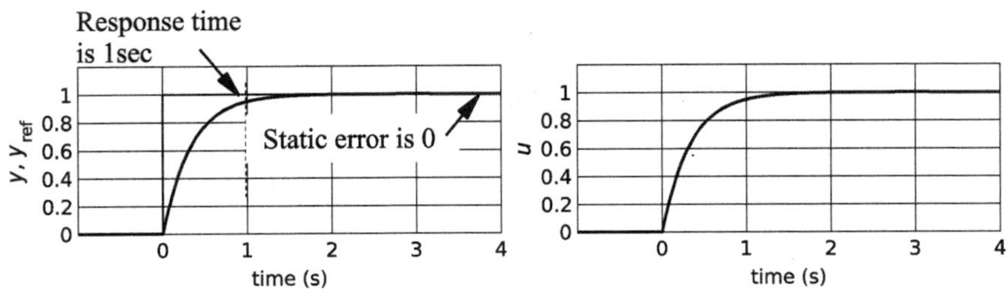

Figure 1.35 Closed-loop response of the system, and controller's output to a step variation of the reference.

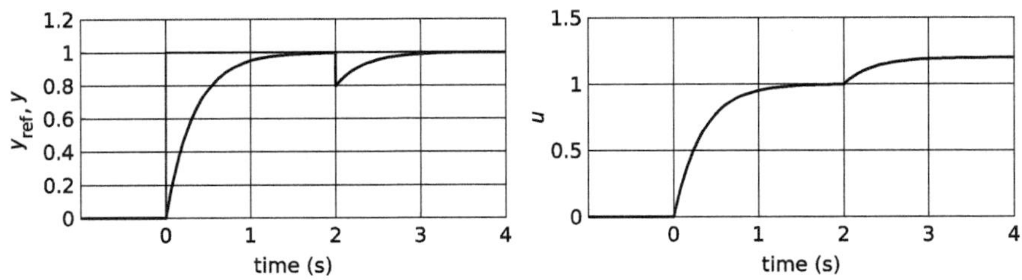

Figure 1.36 Closed-loop responses of the system, and controller's outputs to a step variation of the reference when a perturbation occurs at $t = 2$ s.

Therefore, in a closed loop, the system can be made fast (t_r is small) or slow (t_r is high) simply by changing the value of K_I. In general, K_I is adjusted in order to obtain the response time specified in the requirements, and we calculate

$$K_I = \frac{3}{G \times t_r}. \tag{1.9}$$

We present in Fig. 1.35 the response of the system in closed-loop with $G = 1$ and a required response time $t_r = 1$ s leading to $K_I = 3$. No static error is obtained on system's output, while the 1 s response time is actually obtained.

In simulation, it is easy to check the robustness of the corrector besides perturbations and disturbances. For that purpose, we begin by introducing in the system a perturbation $p(t)$ at $t = 2$ s which consists in a step variation with an amplitude of 0.2, and we plot in Fig. 1.36 the system's response.

A simulation shows that when a perturbation occurs, the system's output instantly drops, creating an error. The error is detected, and automatically, the controller's output increases due to the action of the integral. After 1 s, the system's output recovers its reference, while the input command u has been increased in order to compensate for the perturbation. Hence, the

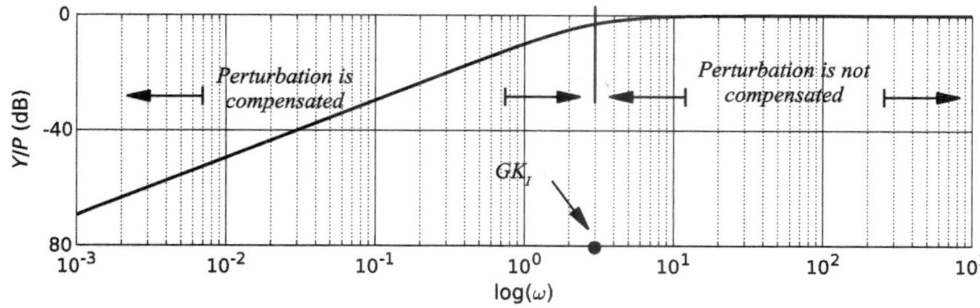

Figure 1.37 Frequency response of the transfer function Y/P: only perturbations with frequency below $K_I G/2\pi$ can be compensated.

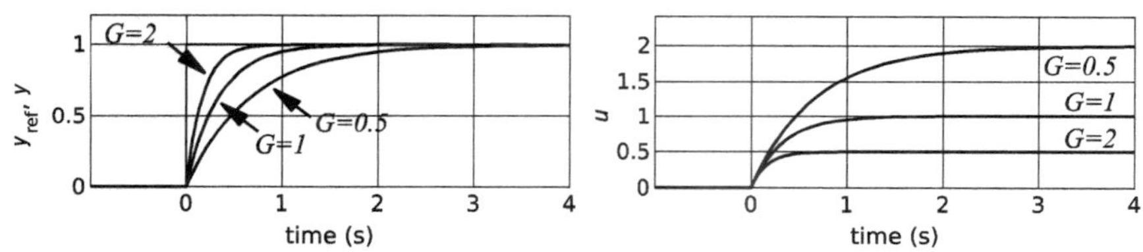

Figure 1.38 Closed-loop responses of the system, and controller's outputs to a step variation of the reference for several values of G.

controller is able to compensate for static perturbations, but not for those which are varying in time. In fact, the transfer function Y/P gives the capability of the closed-loop control to reject perturbations. This transfer function is calculated by

$$\frac{y}{p} = \frac{G}{1+GC} = \frac{G}{1+G\frac{K_I}{s}} = G\frac{\frac{1}{K_I G}s}{1+\frac{1}{K_I G}s}. \tag{1.10}$$

We then plot in Fig. 1.37 the modulus of this transfer function. It shows that for small pulsations (ω is small), Y/P is small, too: a perturbation at these frequencies produces a small variation of y, and the perturbation is considered as rejected. When ω is high, the transfer function is constant, meaning that the controller cannot act to compensate the perturbation: it is too slow, or the perturbation is too fast. The corner frequency is equal to $K_I G/2\pi$.

We consider now the capability of the controller to compensate for the disturbances. The disturbances act on the system's behavior. For example, the disturbances can modify the value of the gain, G, and Fig. 1.38 depicts the system's output when the value of the gain G changes, from $G=1/2$ to $G=2$.

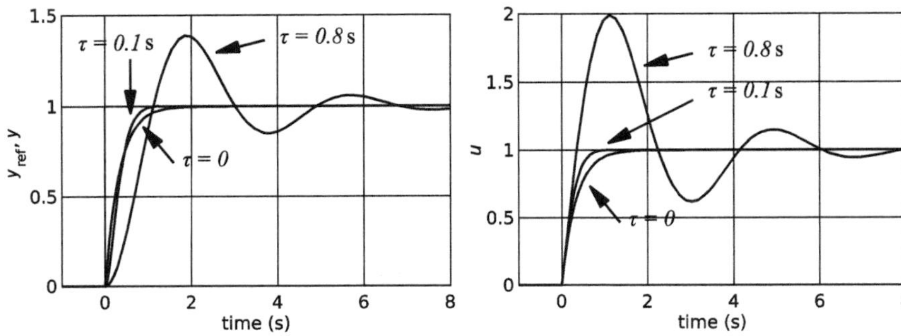

Figure 1.39 System response when taking into account its dominant pole, and comparison with the simple gain model.

This figure shows that while in steady state, the error is null, but during transient period, the system's response is changed: it is faster for $G = 2$ and slower for $G = 1/2$. Hence, the static error is robust to variations of G, but not the response time. Disturbances can also introduce new poles in the system's transfer function $H(s)$. Again, Fig. 1.39 depicts the system's response and the controller's output, when $H(s) = \frac{G}{1+\tau s}$ with $\tau = 0, 0.1, 0.8$ s, the case $\tau = 0$ being equivalent to the simple gain study.

These simulations show that for fast[5] poles, the system's output is almost not modified, while for the slow ones, the system's output becomes oscillating: the closed-loop control destabilizes the system, and therefore, the controller should be changed in order to take into account the slow pole.

Hence, a simple rule of thumb in the controller's design process is that the system's poles should not be taken into account if they are much faster than the closed-loop time constant. If this is not the case, then the closed-loop response time should be increased, or the controller should take into account the slow poles. For this last point, it is convenient to use a PI controller in order to compensate for the slow pole (also called the *dominant* pole).

The system is approximated by its dominant pole: we have $H(s) = \frac{G}{1+\tau s}$. Under that condition, the open-loop transfer function gives

$$OLTF(s) = \left(K_p + \frac{K_I}{s}\right)\frac{G}{1+\tau s} = \frac{GK_I}{s}\frac{1+\frac{K_p}{K_I}s}{1+\tau s}. \tag{1.11}$$

[5] A fast pole corresponds to small values of τ. Indeed, in the frequency domain, it corresponds to a corner frequency $f_c = \frac{1}{2\pi\tau}$. In the same way, slow poles correspond to large values of τ.

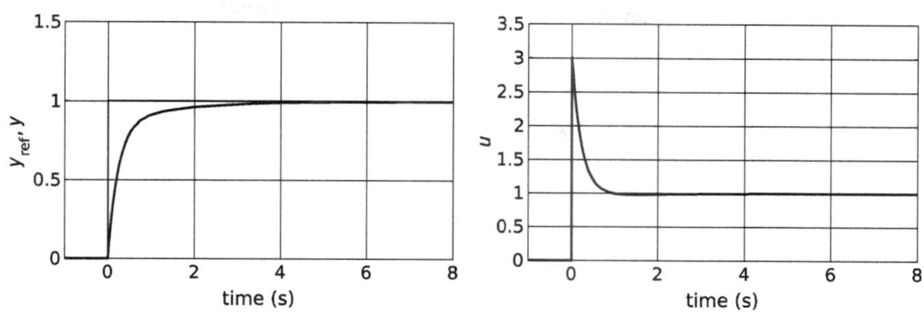

Figure 1.40 System response when taking into account its dominant pole with a PI controller designed in order to have $t_r = 1$ s.

Several methods exist in order to tune the parameters of the PI controllers. One method consists in adjusting K_p and K_I in order to have

$$\frac{K_p}{K_I} = \tau, \tag{1.12}$$

leading to

$$OLTF = \frac{GK_I}{s}. \tag{1.13}$$

Hence, when compensating for the dominant pole, we obtain the same situation as described by Eq. (1.6). Once again, K_I is adjusted in order to have the required closed-loop response time following Eq. (1.7); K_p is deduced from Eq. (1.12), and we have

$$K_p = \frac{3}{G} \times \frac{\tau}{t_r}. \tag{1.14}$$

The system's response y and the controller's output u are depicted in Fig. 1.40 when $G = 1$, $\tau = 0.8$ s, and $t_r = 1$ s, leading to $K_I = 3$ (unchanged) and $K_p = 2.4$; the closed-loop system's response time actually is 1 s without overshoot. Compared with the results depicted in Fig. 1.37, the system's output is the same, but the controller's output u differs during the transient period: it shows a spike at $t = 0$ which accelerates the system's dynamics at the beginning.

We can check controller's robustness to disturbances. If the dominant pole is perfectly compensated, then the conclusion obtained from Fig. 1.37 holds true with a PI controller: the response time is sensitive to the actual value of G. We can also make the pole's value varying, as depicted in Fig. 1.41.

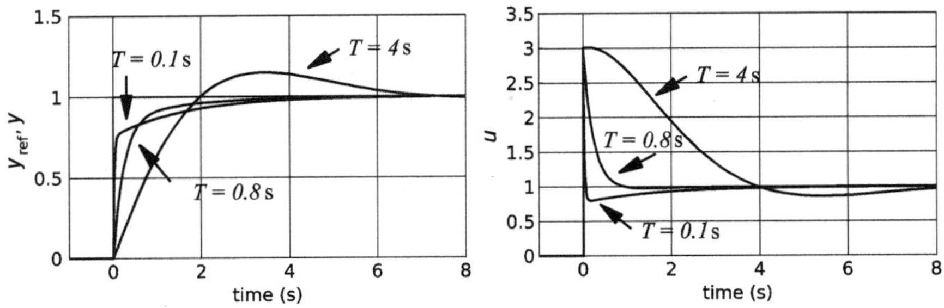

Figure 1.41 System response when taking into account its dominant pole with a PI controller designed in order to have $t_r = 1$ s.

This section describes how it is possible to design a controller for our application. The design is based on a model of the system, and its complexity should correspond to the specifications we set for the system. In particular, the system's poles which fall into the desired closed-loop bandwidth should be taken into account. Faster poles should be let down, for the sake of controller's simplicity.

Also a perfect model is not necessary to attain the requirements, because the controller can mostly compensate the disturbances and the variation of the model's parameters. This is the advantage of the closed-loop control, but this advantage is obtained at the expense of slower dynamics of the system, compared with open-loop operation. Indeed, faster dynamics can also become more unstable as presented above. Therefore, it is essential that the model has a good level of complexity, to take into account enough physical interactions, but simple enough to be able to obtain good performance just by tuning PI controllers. The following chapters will present such models, while the next section presents the state-of-the-art of the control on piezoelectric actuators.

1.4.2 Position control of piezoelectric actuators

The use of feedback controllers to compensate for perturbations and nonlinearity in the piezoelectric actuators has been widely studied. Usually, the voltage is the system's input, but *charge controllers* slightly improve the system's linearity by controlling the charge through the device.

Usually, a simple linear and dynamic model of the actuator as proposed in Eq. (1.4) is used to design the controllers. And when the actuator's stroke is limited, thus showing small hysteresis, a simple PID controller can be used to obtain good performance in a closed loop (Mrad et al., 2003).

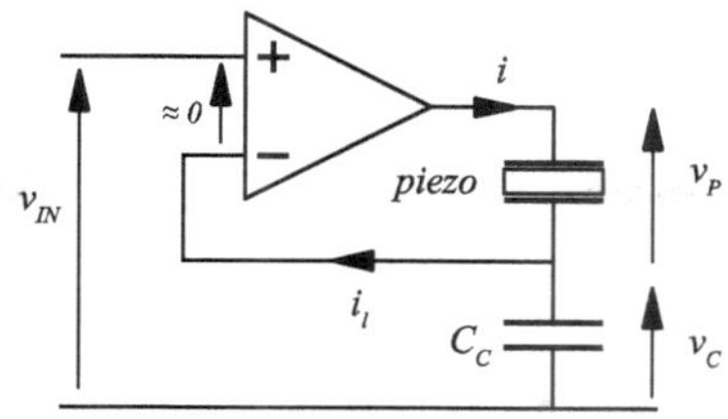

Figure 1.42 A typical charge controller.

Feedforward action can also be used in order to improve the closed-loop system's response time by injecting to the system's input a signal which is directly proportional to the reference setpoint. However, when the load is not known, for example, because its inertia changes with time, or when the stroke is large and the hysteresis is not negligible, this controller shows degraded performance during transient period and steady state error. More advanced controllers are then used, like H_∞ (Ladjal et al., 2009), fuzzy logic (de Abreu and Ribeiro, 2002) and neural network (Ku et al., 2000): improved performances are obtained, but these controllers are harder to implement than the well-known PID controller.

Instead of imposing voltage v across a piezoelectric actuator, one may impose *charge* Q. For a pure and linear dielectric component, charge and voltage are proportional, due to the material's capacitance C, with $Q = C \times v$. For a piece of piezoelectric material, and for no-load operation, the relation between stroke w and charge Q is more linear than the relationship between v and w. Hence, charge controlled actuators are more frequent. A typical charge controller is presented in Fig. 1.42.

The charge controller is based on an operational amplifier (OA). In an ideal OA, the input voltage difference $v_{IN} - v_c$ equals 0. Moreover, the current i_l at the inputs of the OA is also equal to zero. Therefore, the charge flowing in the piezoelectric actuator is equal to the charge in the capacitance C_c, leading to the relationship $Q = C_c v_c$. Since $v_c = v_{IN}$, the charge Q is controlled by the voltage v_{IN} with $Q = C_c v_{IN}$.

In practice the implementation of the charge controller is difficult, because the OA is not perfect, and the current i_l is not null. This limits the performance at low frequency.

1.4.3 Vibration control of ultrasonic actuators

When the actuator is vibrating, which is the case with ultrasonic actuators or piezoelectric motors, the objectives of the control are different than in

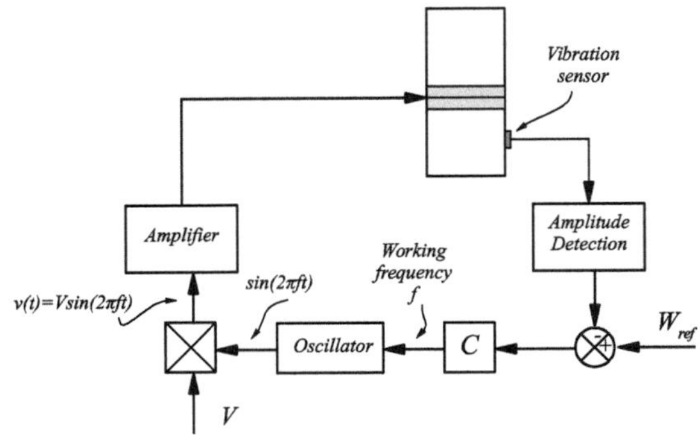

Figure 1.43 Control of the vibration amplitude *W* at constant voltage amplitude *V*, by action on frequency *f*.

the previous section. In fact, the vibration amplitude and its settling time are important. To address this specific issue, a *vibration amplitude control* is achieved.

For most systems, the piezoelectric actuator is supplied with a sinusoidal voltage, at constant amplitude, but with a controllable frequency. The control exploits then the resonance, and, to increase the vibration amplitude, it gets closer to the resonant frequency, while it getting further to reduce it. Hence, the output of the vibration controller is a frequency setpoint, which is the input of an oscillator as detailed in Fig. 1.43.

In this control, the vibration is measured with a sensor, which converts the actuator's displacement w into a voltage. This voltage is sinusoidal; its amplitude is extracted thanks to an *amplitude detection circuit*. The output of the amplitude controller C is the working frequency of an oscillator. After multiplying by a gain V, the signal is then amplified.

One advantage of this control is that it is easy to implement. Moreover, when the actuator's temperature increases and the resonance frequency is decreases, the controller automatically adapts the working frequency in order to maintain a same vibration amplitude as depicted in Fig. 1.44.

However, this control is not optimal, and the disadvantages are twofold:

- First, it will be shown in Chap. 3 that the transfer function between the frequency and vibration amplitude is not linear, and its first pole is not a constant. Therefore, it is necessary to limit the closed-loop response time in order not to destabilize the system. Moreover, the performance strongly depends on the mechanical load of the resonator.
- Second, the system is not operated at resonance. The supplied voltage is then overestimated, compared with an operation at the same vibration

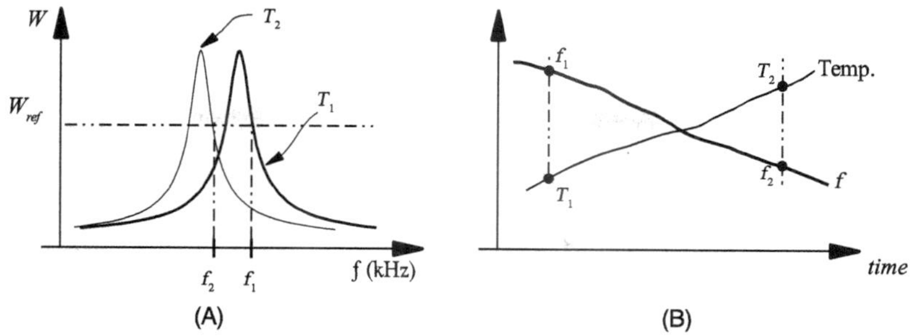

Figure 1.44 Behavior of the frequency control when the temperature changes: (A) evolution of the temperature and the frequency, (B) frequency response of the amplitude *W*.

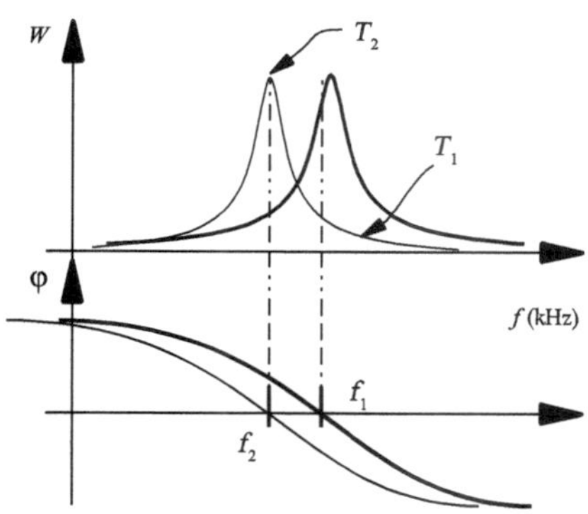

Figure 1.45 Effect of the temperature on the phase shift between *i* and *v*: when the temperature increases, the resonance frequency decreases, but the resonance always corresponds to the same value of the phase shift φ between *i* and *v*.

amplitude at resonance; this leads to higher losses inside the piezoelectric material, and a reduced power efficiency.

Operating at resonance is very interesting because the transfer function is then linear; moreover, the efficiency of the system is increased, compared with operation at a controlled frequency. However, the resonance frequency is not constant in the system, since it varies with the properties of the material used in the system, namely the temperature and load. Therefore, it is necessary to track the resonance frequency, as explained by Fig. 1.45. Most of the *resonance frequency tracking* circuits actually operate at the electrical resonance of the system. Indeed, at the electrical resonance, voltage *v* and current *i* are in phase, and this condition also corresponds to the mechanical resonance when the system is lightly damped.

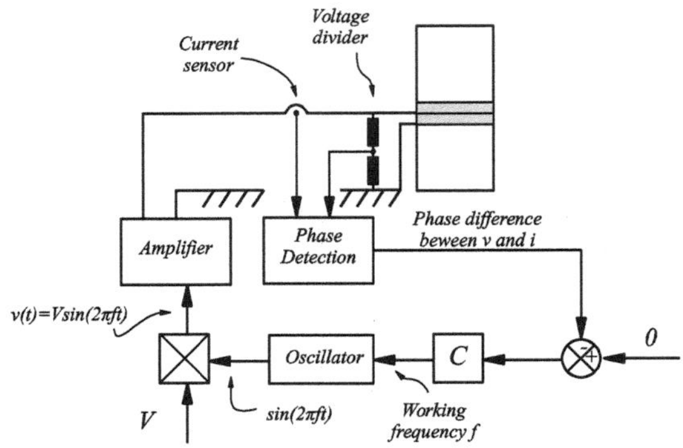

Figure 1.46 A resonance frequency tracking control.

Fig. 1.46 presents an *electrical resonance tracking* circuit, which automatically operates at the electrical resonance. In this circuit, current i and voltage v are measured by two sensors. The phase shift φ between the two signals is then measured thanks to a *phase comparator*. A closed-loop control of the phase is then added in order to adjust the frequency in such a way that the phase equals 0: the voltage and current are in phase, which corresponds to the electrical resonance for the piezoelectric actuator.

When the resonance frequency shifts, due to a temperature rise, for instance, the controller automatically adapts the working frequency, and thus the system still operates at resonance.

The external load can, however, shift the resonance frequency and modify the damping of the system (Takasaki et al., 2007). Therefore, in addition to the resonance frequency tracking system, a parallel closed-loop control will adjust the value of voltage V in order to maintain the right vibration amplitude.

In principle, the tuning of the controller used in these control is not easy, because the transfer function involved in frequency tracking system or the amplitude controller is either nonlinear or nonstationary. In the following chapters, we will present a methodology to design a complete vibration amplitude control of a piezoelectric resonant actuator. This control is similar to the *vector control method*, which has been widely introduced in the control of electromagnetic machines.

CHAPTER TWO

Dynamic modeling of piezoelectric media

Contents

2.1. Introduction 44
2.2. Notations 46
  2.2.1 Repeated indexes summation convention 46
  2.2.2 Derivatives' notation 48
2.3. Elements of elasticity 50
  2.3.1 Strains 50
  2.3.2 Stress 55
  2.3.3 Deformation energy 58
  2.3.4 Elasticity constitutive law 59
2.4. Dynamic equation of an elastic body 60
  2.4.1 Equations of vibrations 60
  2.4.2 Free harmonic vibrations 61
  2.4.3 Orthogonality of mode shapes 64
  2.4.4 Forced vibrations 67
2.5. Dynamic of piezoelectric bodies 74
  2.5.1 Basic of electrostatic 74
    2.5.1.1 Electric field and electric potential 74
    2.5.1.2 Interaction of the electric field with conductive and dielectric media 75
  2.5.2 Piezoelectricity constitutive equations 77
  2.5.3 Electromechanical dynamic equations 79
  2.5.4 Free harmonic equation and orthogonality 80
  2.5.5 Forced vibrations 82
    2.5.5.1 Imposed charges 83
    2.5.5.2 Imposed voltages 84
2.6. Practical considerations 85
  2.6.1 Modeling thin ceramics 85
    2.6.1.1 Longitudinal vibrations 85
    2.6.1.2 Transverse vibrations 87
  2.6.2 Equivalent models 91

Piezoelectric Actuators: Vector Control Method
https://doi.org/10.1016/B978-0-12-814186-1.00006-5

Abstract

The macroscopic dynamic modeling of piezoelectric devices that will be used in the next chapters is presented. The principle of modal decomposition of the dynamic response of elastic media is first introduced. Thanks to this method, the problem is simplified. It leads to simple oscillators accounting for each mode, the actual response is then reconstructed by superposition. This is then extended to piezoelectric structures with the introduction of piezoelectric modal forces. Finally, general models using mechanical or electrical analogies to describe the vibration of the various modes are discussed.

Keywords

macroscopic dynamic modeling, elasticity, electrostatics, piezoelectricity, elastic vibrations, resonance, mode shapes

2.1. Introduction

Piezoelectricity, with its increasing use for various applications such as force or acceleration measurement, echography, ultrasonic actuators, Langevin transducers, radars, atomic force microscopes to cite the more common ones, is a topic that has interested scientists and engineers since its discovery. As a result many models exist, each of them tailored to address specific issues resulting from various points of view. Analytical (Ballas, 2007, Maurini et al., 2006), numerical (Lezgy-Nazargah et al., 2013), equivalent circuits (Sherrit et al., 1999, Mason, 1935), linear, nonlinear (Goldfarb and Celanovic, 1997), quasistatic, dynamic models, all serve a different purpose. Here, the objective is to elaborate a sufficiently precise, yet simple and general, dynamic model for the control. Moreover, it would be better to have some physical insight. The main concepts and techniques are described in the following sections for the reader eager to understand important ideas underpinning the model presented in Sect. 2.6.

Therefore, the approach proposed in this book is to start from the underlying physics. Furthermore, the presentation will be limited to linear laws of piezoelectricity (Ikeda, 1996). Indeed, the piezoelectric coupling in a ferroelectric material is far more complicated. It involves an electromechanical hysteresis which until now can only be modeled by phenomenological approaches. These models are difficult to integrate into a design procedure because they rely on the macroscopic behavior of the whole structure. In other words, they are not predictive in the sense that they require a prior fitting of the model on the existing device. From the control point of view, they are valuable for quasistatic operations where ultraprecise positioning is at stake. This is out of the scope of this book which focuses on the control of vibrating piezoelectric transducers. For these applications, hysteresis induces lag and losses, which can be actually handled in a linear model by introducing an ad-hoc damping (Geradin and Rixen, 2015). Possibly, for high voltage applications, nonlinearities such as amplitude jumps and resonant frequency drift, as exposed in the previous chapter, can occur. These can be handled by the control, again on the basis of a linear model, by adding some adaptive capacity to the controller such as resonant frequency tracking as described in Chap. 1.

Despite the linearity assumption, piezoelectricity remains complex as it combines elasticity and electrostatic concepts. The following material is not exhaustive, this would require too many developments for a single chapter, but it aims at giving the reader keys to understand the main notions that govern the dynamics of piezoelectric media. For this purpose, the essentials of elasticity and electrostatics are recalled in Sects. 2.3 and 2.5.1, respectively.

The topic of elastic vibrations is examined in Sect. 2.4 as it provides the basic concepts for studying piezoelectric media vibrations. An important concept in the study of elastic media vibrations is the resonance that results from the equations which will be presented in Sect. 2.4.1 (Meirovitch, 1967, Geradin and Rixen, 2015). For some specific frequencies that depend on the material properties, the geometry of the structure and the boundary conditions, inertia forces dynamically compensate for the elastic force in the body. This is possible because some special shape of the structure is spontaneously produced by the constructive interference of the vibration reflected by the boundaries. The method to determine those shapes is treated in Sect. 2.4.2.

The importance of these so-called "mode shapes" is central, because they constitute a modal basis on which *any shape in adequation with the kinematic boundary condition* can be decomposed as proved in Sect. 2.4.3. The response to any external loading – which is described by a set of partial differential equations involving time and space derivatives – can thus be solved by decoupling the spatial and temporal dependence of the problem. Briefly put, once the mode shapes are known, space derivatives are eliminated from the problem, resulting in a purely time-dependent model involving an infinite set of linear time-invariant harmonic oscillators. This is the modal decomposition technique described in Sect. 2.4.4.

It should be noted that this modeling is very convenient for the control engineer who is familiar with systems described by time-dependent ordinary differential equations. A second advantage is that modal analysis is a common engineering tool, and so many software packages can actually provide a priori the mode shapes of a given device. Hence the mechanical and control design can be evaluated iteratively in a mechatronic approach.

Section 2.5 then focuses on the dynamics of piezoelectric media. Piezoelectricity was historically considered to be a linear relationship between the elastic and electrostatic quantities (Sect. 2.5.2). This is fairly verified in piezoelectric crystals, which are therefore the preferred materials for precise sensors. It is more of an approximation for ferroelectric ceramics but still holds in a given range of operations. The dynamics of piezoelectric media is thus governed by a set of coupled equations that describes the dynamic vibrations and the conservation of charge, coupled by the piezoelectric material (Sect. 2.5.3). The consequence is twofold:

- With the addition of the dependence on electric quantities, the boundary conditions now include electrical boundary conditions and electrostatic space distribution of forces that are inputs from the control point of view.
- Because of the static nature of the electric part, the dynamic problem is very similar to purely elastic media, consequently the modal decomposition technique is still valid (Sects. 2.5.4 and 2.5.5).

Despite the linearity assumption, the general problem is very complex, and some further simplifications are needed to derive a usable model for the engineer. In Sect. 2.6, the problem is thus limited to thin ceramics where further assumptions can be made regarding the electrical quantities and reduced to one-dimensional fields. This restriction is motivated by the available ceramics that tend to be thinner, and for which the electrodes are designed specifically to impose a dominating field component in a privileged direction. This latest simplification will result in a model that will be applied throughout the rest of the book and eager readers may refer directly to Sect. 2.6.

The various concepts are illustrated by an example at each step of the method for the reader eager to understand the main principles underpinning this chapter, and the following section is dedicated to some conventions and notations that are used in the following developments.

2.2. Notations

2.2.1 Repeated indexes summation convention

In the following sections, many results involve combination and products of tensors and vectors. This leads to complex sums over multiple indexes such as

$$d_j = \sum_{i=1}^{n}\sum_{k=1}^{n} a_{ijk} b_i c_k$$

where n can be any integer. Here, it will be assumed that $(i, j, k) \in \{1, 2, 3\}^3$. After examination of this expression, one can observe that the summations are performed over the indexes that appear two times in the monomial $a_{ijk} b_i c_k$, while the index j does not indicate a summation and refers to the result of the operation yielding three quantities $\{d_1, d_2, d_3\}$. Therefore it appears that the sum sign can be dropped as long as the bound n is known, and as a simplification, we will write

$$d_j = a_{\mathbf{i}j\mathbf{k}} b_{\mathbf{i}} c_{\mathbf{k}}$$

and perform the summation over the *repeated indexes*, which are here highlighted in bold for the sake of explanation. So, the calculations are:

$$\begin{aligned} d_j =& a_{1j1} b_1 c_1 + a_{1j2} b_1 c_2 + a_{1j3} b_1 c_3 + \quad (\text{set } i=1, \text{ iterate over } k) \\ & a_{2j1} b_2 c_1 + a_{2j2} b_2 c_2 + a_{2j3} b_2 c_3 + \quad (\text{set } i=2, \text{ iterate over } k) \\ & a_{3j1} b_3 c_1 + a_{3j2} b_3 c_2 + a_{3j3} b_3 c_3 \quad (\text{set } i=3, \text{ iterate over } k) \end{aligned} \tag{2.1}$$

so each of d_1, d_2, d_3 is the result of the summation of nine monomials. The indexes have different meaning, and so we can make a distinction between a dummy index (i, k in the example) and a real index (j in the same example). A dummy index can be changed without modification of the result, for instance, we can write equivalently

$$d_j = a_{\mathbf{i}j\mathbf{k}} b_{\mathbf{i}} c_{\mathbf{k}} = a_{\mathbf{l}j\mathbf{k}} b_{\mathbf{l}} c_{\mathbf{k}}$$

where the index l is varied on the same range as i. By contrast, j has an actual significance since it addresses three different results, and the coefficients $a_{\mathbf{i}j\mathbf{k}}$ are different for each j considered.

To sum-up, there are two categories of indexes:

1. Summation indexes or dummy indexes must appear *twice (and only twice) on one side of the expression*, and they do not appear on the other side of the equality.
2. Real indexes appear once (and only once) on each side of the equality.

Examples

Consider to matrices $\boldsymbol{A}$ and $\boldsymbol{B}$ with respective dimensions $n \times p$ and $p \times q$ given by

$$\boldsymbol{A} = \begin{bmatrix} a_{11} & a_{12} & \dots & a_{1p} \\ \vdots & & & \vdots \\ a_{n1} & \dots & \dots & a_{np} \end{bmatrix} \quad \text{and} \quad \boldsymbol{B} = \begin{bmatrix} b_{11} & b_{12} & \dots & b_{1q} \\ \vdots & & & \vdots \\ b_{p1} & \dots & \dots & b_{pq} \end{bmatrix},$$

which could, for instance, describe two linear applications. One can compose the two applications in a relevant order to obtain a new application which is described by the matrix $\boldsymbol{C} = \boldsymbol{AB}$. Using the repeated indexes convention (also known as Einstein's convention), each component of $\boldsymbol{C}$ is given by

$$c_{ij} = a_{ik} b_{kj}$$

where $i \in \{1, \dots, n\}$, $j \in \{1, \dots, q\}$ and $k \in \{1, \dots, p\}$.

Consider now the equation

$$u_i = a v_i + b w_i$$

where a and b are scalars. Since i appears only once in each monomial on the right-hand side of the equation, and moreover is also on the left-hand side, there are no summations involved, and it is equivalent to write the vector $\boldsymbol{u}$ as the linear combination of two vectors $\boldsymbol{v}$ and $\boldsymbol{w}$, $\boldsymbol{u} = a\boldsymbol{v} + b\boldsymbol{w}$.

2.2.2 Derivatives' notation

Partial derivatives are often used in the coming sections, and therefore, the following convention is adopted, when unambiguous. Consider an orthonormal basis $\boldsymbol{e}_1$, $\boldsymbol{e}_2$, $\boldsymbol{e}_3$, with the associated coordinates x_1, x_2, x_3, and let $f(x_1, x_2, x_3)$ be a scalar function of the coordinates. Then, one can calculate the partial derivative with respect to one of the coordinates, and we denote it by adding a subscript indicating the derivation direction separated by a comma as follows:

$$\frac{\partial f}{\partial x_1}(x_1, x_2, x_3) = f_{,1}(x_1, x_2, x_3).$$

Higher order derivatives can be described following the same convention, e.g.,

$$f_{,133}(x_1, x_2, x_3) = \frac{\partial^3 f}{\partial x_1 \partial {x_3}^2}(x_1, x_2, x_3).$$

This notation can be combined with the previous one. For instance, let $\boldsymbol{v}$ be a vector field of space, defined as the gradient of the function f. It can be denoted in two ways:

$$\boldsymbol{v}(x_1, x_2, x_3) = \nabla f(x_1, x_2, x_3) = \begin{bmatrix} \frac{\partial f}{\partial x_1}(x_1, x_2, x_3) \\ \frac{\partial f}{\partial x_2}(x_1, x_2, x_3) \\ \frac{\partial f}{\partial x_3}(x_1, x_2, x_3) \end{bmatrix}$$
$$\Leftrightarrow v_i(x_1, x_2, x_3) = f_{,i}(x_1, x_2, x_3).$$

Here, index $i \in \{1, 2, 3\}$ describes the three components of the gradient vector as it appears on both sides of the equality. It is a real index, and the comma is here to avoid the confusion between summation indexes and derivative direction, but they are iterated simultaneously. Hence, this notation is consistent with the Einstein's convention rules enunciated above.

Suppose now that the divergence of $\boldsymbol{v}$ is to be computed. It is a scalar given by (the coordinates are omitted here)

$$d = \frac{\partial v_1}{\partial x_1} + \frac{\partial v_2}{\partial x_2} + \frac{\partial v_3}{\partial x_3} \Leftrightarrow d = v_{i,i} = f_{,ii}.$$

Since there are no indexes on the left-hand side, i is regarded as a dummy index which implies the summation. We conclude by the following example:

$$T_{ij,j} = 0.$$

Here, the result is seemingly a scalar, however, as the index is i isolated, it indicates that this equation actually describes three different equations. So it should be understood as $T_{ij,j} = a_i$, $a_i = 0$, that is,

$$T_{i1,1} + T_{i2,2} + T_{i3,3} = 0 \quad \text{for} \quad i \in \{1, 2, 3\}.$$

To conclude this part, time derivative will often be denoted by the "dot" notation. For instance, the first and second time derivatives of a function $g(t)$ are written:

$$\frac{dg(t)}{dt} = \dot{g}(t) \text{ and } \frac{d^2 g(t)}{dt^2} = \ddot{g}(t).$$

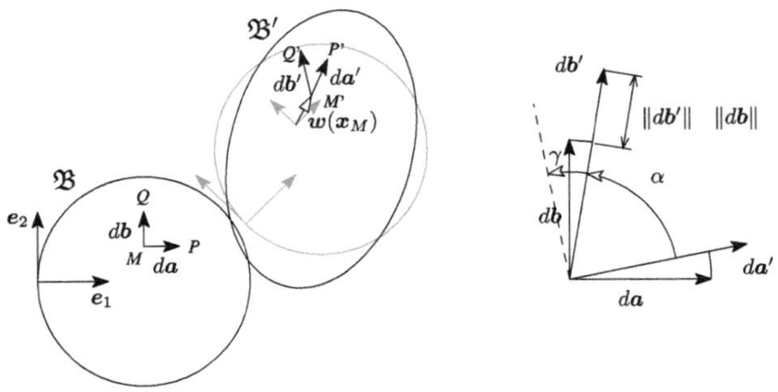

Figure 2.1 Statement of the elastic problem. Strains measure local change in length of infinitesimal vectors, and modification of angle between two infinitesimal vectors.

2.3. Elements of elasticity

2.3.1 Strains

A body subjected to forces can be simultaneously displaced and deformed as depicted in Fig. 2.1. The initial configuration is called the *reference configuration* of the body (here denoted by $\mathfrak{B}$). First, a displacement leaves the body intrinsically unchanged but its position in space, described in a given frame, is modified; this is called a rigid body motion. The body can be translated and rotated, but its shape remains; in the figure, this is outlined by the grayed shape.

A deformation, on the other hand, modifies shapes, which is the reason why in the figure the deformed configuration $\mathfrak{B}'$ differs from the reference configuration. The "warping" of the body can be defined intuitively by the modification of length between two points before and after deformation, or the change in the angle between two vectors defined at the same point (Geradin and Rixen, 2015). To measure this, three perpendicular infinitesimal vectors $\boldsymbol{da} = \boldsymbol{MP}$, $\boldsymbol{db} = \boldsymbol{MQ}$ and $\boldsymbol{dc} = \boldsymbol{MR}$ are defined at a point M on the reference configuration as depicted (cf. Fig. 2.1 where for the sake of clarity only two dimensions are represented). The deformation moves these points to M', P', Q', and so on. We define the displacement field, a vector function of coordinates $\boldsymbol{w}(x_1, x_2, x_3)$ that measures the displacement of a material point in the body *relative to its position in the reference configuration.* So, if the point M is initially located at $\boldsymbol{OM} = [x_{1M}, x_{2M}, x_{3M}]^\top = \boldsymbol{x}_M$, then the new position is

$$\begin{bmatrix} x'_{1M} \\ x'_{2M} \\ x'_{3M} \end{bmatrix} = \begin{bmatrix} x_{1M} \\ x_{2M} \\ x_{3M} \end{bmatrix} + \begin{bmatrix} w_1(x_{1M}, x_{2M}, x_{3M}) \\ w_2(x_{1M}, x_{2M}, x_{3M}) \\ w_3(x_{1M}, x_{2M}, x_{3M}) \end{bmatrix} = \boldsymbol{x}_M + \boldsymbol{w}(\boldsymbol{x}_M).$$

Note that the displacement field is defined as a function of the coordinates of the points *of the reference configuration*. First, we consider the vector $\boldsymbol{da}$, and calculate the "warped" vector $\boldsymbol{da}'$ as

$$\boldsymbol{da}' = \boldsymbol{x}'_P - \boldsymbol{x}'_M = \boldsymbol{x}_P + \boldsymbol{w}(x_P) - \boldsymbol{x}_M + \boldsymbol{w}(x_M) = \boldsymbol{da} + \boldsymbol{w}(\boldsymbol{x}_M + \boldsymbol{da}) - \boldsymbol{w}(\boldsymbol{x}_M). \tag{2.2}$$

Considering a first-order approximation, one has

$$\boldsymbol{w}(\boldsymbol{x}_M + \boldsymbol{da}) = \boldsymbol{w}(\boldsymbol{x}_M) + \nabla \boldsymbol{w}(\boldsymbol{x}_M)\, \boldsymbol{da} + \mathrm{o}(\|\boldsymbol{da}\|^2), \tag{2.3}$$

that is, if the previous expression is rewritten componentwise, we get

$$w_i(\boldsymbol{x}_M + \boldsymbol{da}) = w_i(\boldsymbol{x}_M) + w_{i,j}(\boldsymbol{x}_M)\, da_j + \mathrm{o}(a_k a_k). \tag{2.4}$$

Substituting Eq. (2.4) into Eq. (2.2), and neglecting the higher-order terms, one gets the following approximation for the components of the vector $\boldsymbol{da}'$:

$$da'_i = da_i + w_{i,j}(\boldsymbol{x}_M)\, da_j. \tag{2.5}$$

Let $\boldsymbol{db}$ be another vector defined at the same point M. We intend to measure how the scalar product of $\boldsymbol{da}$ and $\boldsymbol{db}$ is modified by the deformation. For this purpose, using the previous result, one gets

$$da'_i db'_i = \left(da_i + w_{i,j}(\boldsymbol{x}_M)\, da_j\right)\left(db_i + w_{i,j}(\boldsymbol{x}_M)\, db_j\right). \tag{2.6}$$

After expanding the product, the equation reads

$$da'_i db'_i = da_i db_i + 2S_{ij} da_i db_j \tag{2.7}$$

where the strain components S_{ij} have been introduced. They are given by

$$2S_{ij} = w_{i,j} + w_{i,j} + w_{k,i} w_{k,j}. \tag{2.8}$$

In the scope of this book, the local displacement will be considered to be small, that is, only small perturbations with respect to the reference configuration will be considered. In other words, the warping is small, and therefore the *linearized strains* are considered:

$$2S_{ij} = w_{i,j} + w_{i,j}, \tag{2.9}$$

i.e., the strains are approximated to a first order. The components of the strain are symmetrical; indeed, one can easily verify that $S_{ij} = S_{ji}$.

To measure the *relative* change of length in one direction, the vector $\boldsymbol{da}$ (see Fig. 2.1) is considered, and the squared length of $\boldsymbol{da}'$ is deduced using Eq. (2.7) as

$$\|\boldsymbol{da}'\|^2 = \left(da_i + w_{i,j}(\boldsymbol{x}_M)\, da_j\right)^2 = da_i^2 + 2S_{ij}da_i da_j, \tag{2.10}$$

which is used to define the relative length variation

$$\begin{aligned}\frac{\|\boldsymbol{da}'\| - \|\boldsymbol{da}\|}{\|\boldsymbol{da}\|} &= \frac{\sqrt{da_i\delta_{ij}da_j + 2da_iS_{ij}da_j} - \|\boldsymbol{da}\|}{\|\boldsymbol{da}\|} = \\ &= \sqrt{\frac{da_i^2}{\|\boldsymbol{da}\|^2} + 2\frac{da_i}{\|\boldsymbol{da}\|}S_{ij}\frac{da_j}{\|\boldsymbol{da}\|}} - 1.\end{aligned} \tag{2.11}$$

Here, δ_{ij} is the Kronecker symbol, which is equal to 1 if $i = j$ and 0 otherwise. Introducing the unit vector $\boldsymbol{v}$ in the direction of $\boldsymbol{da}$, that is, $\boldsymbol{da} = \|\boldsymbol{da}\|\boldsymbol{v}$, one gets

$$\frac{\|\boldsymbol{da}'\| - \|\boldsymbol{da}\|}{\|\boldsymbol{da}\|} = \sqrt{1 + v_iS_{ij}v_j} - 1. \tag{2.12}$$

So, in the case of small strains, the relative variation of length of $\boldsymbol{da}$ is

$$\frac{\|\boldsymbol{da}'\| - \|\boldsymbol{da}\|}{\|\boldsymbol{da}\|} \simeq v_iS_{ij}v_j, \tag{2.13}$$

which results from a first order approximation of Eq. (2.12).

The two vectors $\boldsymbol{da}$ and $\boldsymbol{db}$ of the figure are originally orthogonal, and form a different angle after deformation. It can be measured by (cf. Fig. 2.1)

$$\cos(\alpha) = \frac{\boldsymbol{da}' \cdot \boldsymbol{db}'}{\|\boldsymbol{da}'\| \times \|\boldsymbol{db}'\|} = \sin(\gamma). \tag{2.14}$$

So, according to Eq. (2.7), one can rewrite the equation as

$$\sin\gamma = \frac{\boldsymbol{da} \cdot \boldsymbol{db} + 2\|\boldsymbol{da}\| \times \|\boldsymbol{db}\| \times \left(v_iS_{ij}z_j\right)}{\|\boldsymbol{da}'\| \times \|\boldsymbol{db}'\|} \tag{2.15}$$

where z_j are the components of the unit vector in direction $\boldsymbol{db}$. Since the vectors $\boldsymbol{da}$ and $\boldsymbol{db}$ are orthogonal, their scalar product is zero. Moreover, the deformations are supposed to be small, so the first-order approximation gives the following simplifications: $\|\boldsymbol{db}'\| = \|\boldsymbol{db}\| \times (1 + z_i S_{ij} z_j)$, $\|\boldsymbol{da}'\| = \|\boldsymbol{da}\| \times (1 + v_i S_{ij} v_j)$, and $\sin\gamma \simeq \gamma$. Finally, in the frame of small perturbations, the angle between two orthogonal vectors in the neighborhood of point M is given by

$$\gamma = \frac{2 v_i S_{ij} z_j}{(1 + v_i S_{ij} v_j) \times (1 + z_i S_{ij} z_j)}. \tag{2.16}$$

Example 1 Some simple strain tensors

For a better understanding of the meaning of the strains, we study a rectangular parallelepiped subject to various plane strain states (so only two dimensions have to be considered).

First, let $\boldsymbol{w}$, the displacement field, be such that $S_{12} = 0$. For instance, it is easy to verify that this holds for

$$\boldsymbol{w} = \begin{bmatrix} k_1 x_1 \\ k_2 x_2 \end{bmatrix}$$

for some real values k_1 and k_2. The strains are then written using Eq. (2.9) as

$$2S_{11} = 2k_1, \quad 2S_{22} = 2k_2, \quad S_{12} = 0$$

as expected. Such a field is represented in Fig. 2.2 (A). The vectors $\boldsymbol{a}$ and $\boldsymbol{b}$ that will be discussed hereafter could be any concomitant perpendicular edges of the grayed grid: it is clear that their lengths are modified but they remain perpendicular. Moreover, since the strains do not depend on the position, the deformation is called uniform. The relative change of length of an horizontal vector $\boldsymbol{da} = \|\boldsymbol{da}\| \times [1\ 0]^{\top}$ in the body is thus given by

$$\frac{\|\boldsymbol{da}'\| - \|\boldsymbol{da}\|}{\|\boldsymbol{da}\|} = (1 \times k_1 \times 1 + 0 \times k_2 \times 0)\,\|\boldsymbol{da}\| = k_1 \|\boldsymbol{da}\|$$

and, when considering the orthogonal vector $\boldsymbol{db} = \|\boldsymbol{db}\| \times [1\,0]^{\top}$, one gets $\frac{\|\boldsymbol{db}'\| - \|\boldsymbol{db}\|}{\|\boldsymbol{db}\|} = (0 \times k_1 \times 0 + 1 \times k_2 \times 2)\|\boldsymbol{db}\| = k_2 \|\boldsymbol{db}\|$. The reader can also easily verify that $\gamma = 0$.

Now, we examine the case of a uniform displacement field where only $S_{12} \neq 0$. To verify $S_{11} = w_{1,1}$ and $S_{22} = w_{2,2}$, the components w_1 and w_2 should depend respectively only on x_2 and x_1. Therefore, $w_{1,2} + w_{2,1} = k$ where k is some constant to verify the uniformity condition. Such a field looks approximately as depicted in Figs. 2.2 (B) and (C). It is given by

$$w = \begin{bmatrix} k_1 x_2 \\ k_2 x_1 \end{bmatrix}$$

and then

$$S_{11} = 0, \quad S_{22} = 0, \quad 2S_{12} = k_1 + k_2 = k.$$

This implies for the vectors da and db that

$$\|da'\| = \|da\|, \quad \|db'\| = \|db\|.$$

In Fig. 2.2 (B), examples of vectors experiencing pure shear strains da, db are sketched. The new vectors resulting from the deformation da', db' have not changed in length, as the dashed–dotted blue circles (dark gray dashed–dotted circles in print version) show : clearly, the two vectors are still radii of the translated circle. However, their relative position is modified: this is visible considering the angle between the vectors before and after deformation, namely

$$\gamma = 2 \times 1 \times S_{12} \times 1 = k. \tag{2.17}$$

Suppose now that da and db have the same modulus (as actually depicted in the figure). Consider the two vectors $dc = \frac{1}{\sqrt{(2)}}(da + db)$ and $dd = \frac{1}{\sqrt{(2)}}(-da + db)$ which are two orthogonal vectors having the same moduli as da and db and forming an angle of $\frac{\pi}{4}$ with them (see Fig. 2.2 (C)). Applying Eq. (2.13), one finds that the relative variations in length are $\frac{\gamma}{2}\|da\|$ (dilatation) and $-\frac{\gamma}{2}\|da\|$ (compression) for dc and dd, respectively, while the angle between them is conserved. This is also sketched in the figure using a circle to materialize their respective length. The example reveals that the strain measures deformations which depend on the direction considered. In other words, if we had chosen a reference frame aligned with c and d, the strain component S_{12} would be zero. It also illustrates that it is always possible to find an orientation such that the components $S_{ij} = 0$ for $i \neq j$.

To conclude this example, we consider a special case of rigid body motion where all strains are null. From the previous calculation, this is possible if $k_1 + k_2 = 0$ and the displacement field is then

$$w = \begin{bmatrix} -k_1 x_2 \\ k_1 x_1 \end{bmatrix} \tag{2.18}$$

which, as depicted in the same figure on the right, is a small local rotation of the domain.

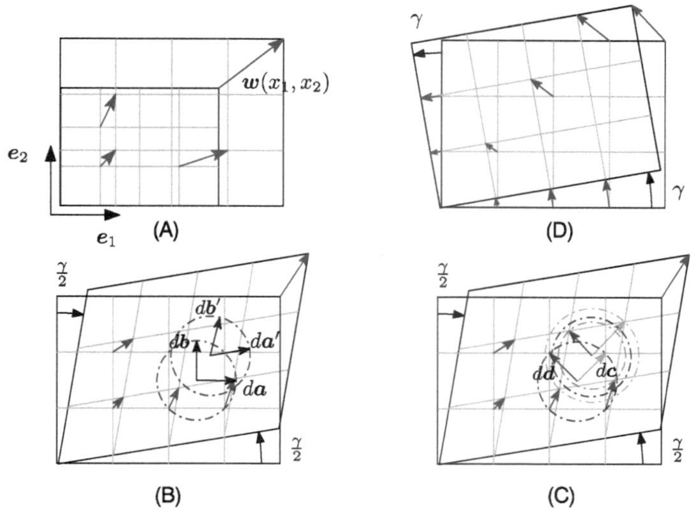

Figure 2.2 Example of basic displacements field: (A) a deformation involving no shear strain in the depicted reference frame. The field at the bottom is a pure shear strain when expressed in the reference frame; however, strains differ according to the directions considered (B) and (C). In (D) the body is rotated, yielding no strains (rigid body displacement).

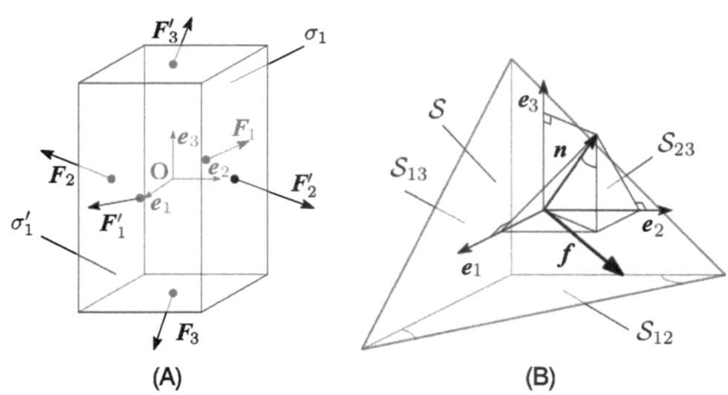

Figure 2.3 Loaded infinitesimal geometries used to derive the stress properties.

2.3.2 Stress

An elastic body opposes reaction forces to a deformation that is caused by an external mechanical loading. A classical assumption is that only forces (and no moments) can be applied at an infinitesimal scale (Spencer, 2012), which leads to the introduction of stresses as the force components applied to an elementary surface. We shall denote by T_{ij} the components of the stress tensor in direction i. We consider now a very small rectangular parallelepiped and derive the static equilibrium equations. The axes are depicted in the figure, and $\boldsymbol{F}_i$ denotes the forces applied along direction i facing the negative side of the axis, while $\boldsymbol{F}'_i$ is used for the opposite face. According to the definition of the stress, the force on $\mathcal{S}_1$ (cf. Fig. 2.3 (A)) is

$$\boldsymbol{F}_1 = \begin{bmatrix} -T_{11}\,dx_2\,dx_3 \\ -T_{21}\,dx_2\,dx_3 \\ -T_{31}\,dx_2\,dx_3 \end{bmatrix}, \tag{2.19}$$

and on the opposite face, the force has slightly changed, so, to a first order,

$$\boldsymbol{F}'_1 = \begin{bmatrix} (T_{11}+T_{11,1}\,dx_1)\,dx_2\,dx_3 \\ (T_{21}+T_{21,1}\,dx_1)\,dx_2\,dx_3 \\ (T_{31}+T_{31,1}\,dx_1)\,dx_2\,dx_3 \end{bmatrix}. \tag{2.20}$$

The same process is applied to express the other forces, and one gets:

$$\boldsymbol{F}_2 = \begin{bmatrix} -T_{12}\,dx_1\,dx_3 \\ -T_{22}\,dx_1\,dx_3 \\ -T_{32}\,dx_1\,dx_3 \end{bmatrix}, \quad \boldsymbol{F}'_2 = \begin{bmatrix} (T_{12}+T_{12,2}\,dx_2)\,dx_1\,dx_3 \\ (T_{22}+T_{22,2}\,dx_2)\,dx_1\,dx_3 \\ (T_{32}+T_{32,2}\,dx_2)\,dx_1\,dx_3 \end{bmatrix}, \quad \boldsymbol{F}_3 = \begin{bmatrix} -T_{13}\,dx_1\,dx_2 \\ -T_{23}\,dx_1\,dx_2 \\ -T_{33}\,dx_1\,dx_2 \end{bmatrix},$$

$$\boldsymbol{F}'_3 = \begin{bmatrix} (T_{13}+T_{13,3}\,dx_3)\,dx_1\,dx_2 \\ (T_{23}+T_{23,3}\,dx_3)\,dx_1\,dx_2 \\ (T_{33}+T_{33,3}\,dx_3)\,dx_1\,dx_2 \end{bmatrix}. \tag{2.21}$$

Additionally, volume forces, such as gravity, for instance, can be applied to the elementary volume, which can be written as

$$\boldsymbol{P} = \begin{bmatrix} P_1 \\ P_2 \\ P_3 \end{bmatrix}. \tag{2.22}$$

At static equilibrium, one has

$$\boldsymbol{F}_1 + \boldsymbol{F}'_1 + \boldsymbol{F}_2 + \boldsymbol{F}'_2 + \boldsymbol{F}_3 + \boldsymbol{F}'_3 + \boldsymbol{P} = 0. \tag{2.23}$$

So, once the forces are replaced using the relations above, the equilibrium is found in terms of the stress components as follows:

$$\left\{ \begin{array}{lcl} T_{11,1} + T_{12,2} + T_{13,3} + P_1 & = & 0 \\ T_{21,1} + T_{22,2} + T_{23,3} + P_2 & = & 0 \\ T_{31,1} + T_{32,2} + T_{33,3} + P_3 & = & 0 \end{array} \right. \Leftrightarrow T_{ij,j} + P_i = 0. \tag{2.24}$$

The elementary volume is also in equilibrium with respect to the moment. Consider the moment applied to the faces perpendicular to the $\boldsymbol{e}_1$ direction.

They are given by

$$\boldsymbol{M}_{F_1} = \mathbf{OQ_1} \wedge \boldsymbol{F_1} = \begin{vmatrix} -\frac{dx_1}{2} & & -T_{11}\,dx_2dx_3 \\ 0 & \wedge & -T_{21}\,dx_2dx_3 \\ 0 & & -T_{31}\,dx_2dx_3 \end{vmatrix} = \begin{bmatrix} 0 \\ -T_{31}\frac{d\Omega}{2} \\ T_{21}\frac{d\Omega}{2} \end{bmatrix} \tag{2.25}$$

and

$$\boldsymbol{M}_{F_1'} = \mathbf{OQ_1'} \wedge \boldsymbol{F_1} = \begin{vmatrix} \frac{dx_1}{2} & & (T_{11}+T_{11,1}dx_1)dx_2dx_3 \\ 0 & \wedge & (T_{21}+T_{21,1}dx_1)dx_2dx_3 \\ 0 & & (T_{31}+T_{31,1}dx_1)dx_2dx_3 \end{vmatrix} = \begin{bmatrix} 0 \\ -(T_{31}+T_{31,1}dx_1)\frac{d\Omega}{2} \\ (T_{21}+T_{21,1}dx_1)\frac{d\Omega}{2} \end{bmatrix} \tag{2.26}$$

where $d\Omega = dx_1dx_2dx_3$ is the elementary volume. Iterating the same process over the other directions gives:

$$\boldsymbol{M}_{F_2} = \begin{bmatrix} T_{32}\frac{d\Omega}{2} \\ 0 \\ -T_{12}\frac{d\Omega}{2} \end{bmatrix}, \ \boldsymbol{M}_{F_2'} = \begin{bmatrix} (T_{32}+T_{32,2}dx_2)\frac{d\Omega}{2} \\ 0 \\ -(T_{12}+T_{12,2}dx_2)\frac{d\Omega}{2} \end{bmatrix}, \ \boldsymbol{M}_{F_3} = \begin{bmatrix} -T_{23}\frac{d\Omega}{2} \\ T_{13}\frac{d\Omega}{2} \\ 0 \end{bmatrix},$$
$$\boldsymbol{M}_{F_3'} = \begin{bmatrix} -(T_{23}+T_{23,3}dx_3)\frac{d\Omega}{2} \\ (T_{13}+T_{13,3}dx_3)\frac{d\Omega}{2} \\ 0 \end{bmatrix}, \tag{2.27}$$

and since the volume force is applied at the center of the elementary volume, it produces no moment. Writing the second Newton law at first order yields

$$\begin{cases} T_{32}d\Omega - T_{23}d\Omega &= 0, \\ -T_{31}d\Omega + T_{13}d\Omega &= 0, \\ T_{21}d\Omega - T_{12}d\Omega &= 0. \end{cases} \tag{2.28}$$

This is realized if $T_{12} = T_{21}$, $T_{23} = T_{32}$ and $T_{31} = T_{13}$. In other words, the stress tensor is also symmetric.

To relate the stress tensor to the external forces, the equilibrium of an infinitesimal tetrahedron (cf. Fig. 2.3 (B)) is now studied. Three of its edges have the same infinitesimal length and are parallel to the axes. They form the three faces denoted $\mathcal{S}_{13}$, $\mathcal{S}_{23}$ and $\mathcal{S}_{12}$ normal to the reference frame directions $\boldsymbol{e}_1$, $\boldsymbol{e}_2$ and $\boldsymbol{e}_3$, respectively. The remaining face is $\mathcal{S}$, and it is subjected to a force surface density $\boldsymbol{\tau}$, resulting in the force $\boldsymbol{f} = \boldsymbol{\tau}\mathcal{S}$. Let α, β, γ be the angles formed by $\boldsymbol{n}$, the outward normal of $\mathcal{S}$ with the reference frame directions. The forces applied on $\mathcal{S}_{13}$, $\mathcal{S}_{23}$ and $\mathcal{S}_{12}$ are:

$$\boldsymbol{F}_1 = \begin{bmatrix} T_{11}\mathcal{S}_{23} \\ T_{21}\mathcal{S}_{23} \\ T_{31}\mathcal{S}_{23} \end{bmatrix}, \ \boldsymbol{F}_2 = \begin{bmatrix} T_{12}\mathcal{S}_{13} \\ T_{22}\mathcal{S}_{13} \\ T_{32}\mathcal{S}_{13} \end{bmatrix}, \ \boldsymbol{F}_3 = \begin{bmatrix} T_{13}\mathcal{S}_{12} \\ T_{23}\mathcal{S}_{12} \\ T_{33}\mathcal{S}_{12} \end{bmatrix}. \tag{2.29}$$

The various surfaces involved above are related to $\mathcal{S}$ by the projection $\mathcal{S}_{23} = \mathcal{S}\cos\alpha$, $\mathcal{S}_{13} = \mathcal{S}\cos\beta$, $\mathcal{S}_{12} = \mathcal{S}\cos\gamma$, so writing the equilibrium and simplifying by S leads to

$$\begin{cases} T_{11}\cos\alpha + T_{12}\cos\beta + T_{13}\cos\gamma &= \tau_1, \\ T_{21}\cos\alpha + T_{22}\cos\beta + T_{23}\cos\gamma &= \tau_2, \\ T_{31}\cos\alpha + T_{32}\cos\beta + T_{33}\cos\gamma &= \tau_3. \end{cases}$$

Since the cosine actually represents the coordinate of the normal vector $\boldsymbol{n}$, the expression above can be rewritten in a compact form as

$$T_{ij}n_j = \tau_i. \tag{2.30}$$

2.3.3 Deformation energy

A body is deformed by the application of an external loading, and the resulting displacement field respects the kinematic constraints that depend on how the body of interest is connected to its environment. To isolate the deformation energy, it is assumed that the only forces applied are due to the force distribution $\underline{\tau}$ on the boundary Σ of the body. To disturb the deformed body by a small amount near its equilibrium position, the surrounding environment must provide an increment of energy $\delta\mathcal{W}_{ex}$:

$$\delta\mathcal{W}_{ex} = \iint_{\Sigma} \underline{\tau}_i \delta w_i d\Sigma.$$

Owing to Eq. (2.30), the previous equality becomes

$$\delta\mathcal{W}_{ex} = \iint_{\Sigma} \delta w_i T_{ij} n_j d\Sigma.$$

To make the connection with the deformation state inside the body, the divergence theorem is recalled: supposing that Ω is compact and Σ its boundary is piecewise continuous, for a smooth vector field $\boldsymbol{V}$ we have

$$\iiint_{\Omega} V_{i,i} d\Omega = \iint_{\Sigma} V_i n_i d\Sigma, \tag{2.31}$$

which in the current case results in

$$\iint_{\Sigma} \delta w_i T_{ij} n_j d\Sigma = \iiint_{\Omega} \left(\delta w_i T_{ij}\right)_{,j} d\Omega = \iiint_{\Omega} \left(\delta w_{i,j} T_{ij} + \delta w_i T_{ij,j}\right) d\Omega$$

because $\delta w_i T_{ij}$ actually define the components of a vector field. Using the equilibrium Eq. (2.24) for $\boldsymbol{P} = 0$, one can eliminate the second term under the volume integral sign, and finally reordering the sum so as to make the definition of strain (2.9) appear, one finally gets

$$\delta W_{ex} = \iiint_{\Omega} T_{ij} \delta S_{ij} d\Omega .$$

So, the elementary increment of the energy stored by deformation is the volume integral of the local deformation energy:

$$\delta w_e = T_{ij} \delta S_{ij}. \tag{2.32}$$

2.3.4 Elasticity constitutive law

An elastic body opposes forces to external actions that induce its deformation. Once the cause of the deformation disappears, the body returns to its reference configuration, while stresses and strains vanish (assuming that the reference configuration involves no residual stresses which are beyond the scope of this book). The stresses result from the strain inside the body during deformation, and it remains to define how they are related. Here, the discussion is limited to *linear* elasticity, that is, we suppose that the stresses vary linearly with the strains. Under an increment of strains dS_{ij}, the increment of stresses is therefore written as

$$dT_{ij} = c_{ijkl} dS_{kl}. \tag{2.33}$$

The c_{ijkl} are constants, called the stiffness tensor coefficients, which satisfy the symmetries: $c_{ijkl} = c_{jikl}$, $c_{ijkl} = c_{ijlk}$ and $c_{ijkl} = c_{klij}$ (i.e., there are at most 21 independent stiffness coefficients). Equation (2.33) is the differential form of the linear elastic constitutive law,

$$T_{ij} = c_{ijkl} S_{kl}. \tag{2.34}$$

Alternatively, the elastic law can relate the strains to stresses using the compliance tensor coefficients as follows:

$$S_{ij} = s_{ijkl} T_{kl}. \tag{2.35}$$

Substituting in (2.32) the law (2.34), one can express the energy density at a given deformation state, denoted by A, by integrating the energy variation from rest to the considered state:

$$w_E(A) = \int_0^A \delta w_E dS = \int_0^A c_{ijkl} S_{kl} dS_{ij} = \frac{1}{2} c_{ijkl} S_{ij}(A) S_{kl(A)}$$

$$\Leftrightarrow w_E(A) = \frac{1}{2} T_{ij}(A) S_{ij}(A). \tag{2.36}$$

The energy does not depend on the way the components of the strains are varied to reach state A but only on their values for this point: the deformation energy is a state function.

Van Voigt notations

The stress and strain are symmetric, so each tensor only has six independent coordinates. The tensor notation is useful for demonstrating some important properties that will be examined later, however, it is sometimes cumbersome. Later, in some examples, the following simplification will be adopted: $\{T_{11}, T_{22}, T_{33}, T_{23}, T_{13}, T_{12}\} = \{T_1, T_2, T_3, T_4, T_5, T_6\}$ and $\{S_{11}, S_{22}, S_{33}, 2S_{23}, 2S_{13}, 2S_{12}\} = \{S_1, S_2, S_3, S_4, S_5, S_6\}$. This results in a vector notation for the stress and strain. Consequently, the elasticity tensors are represented by 6×6 matrices, which is possible due to its symmetries. To do so, the corresponding rule is applied: $11 \rightarrow 1$, $22 \rightarrow 2$, $33 \rightarrow 3$, 12 and $21 \rightarrow 6$, 23 and $32 \rightarrow 5$, 31 and $13 \rightarrow 6$. For instance, the stiffness $c_{1123} = c_{15}$.

Finally, the coefficients that were introduced in the definition of the strain vector components in relation to the strain tensor components ensure that the energy expression remains unchanged, that is,

$$w_E = \frac{1}{2} T_{ij} S_{ij} = \frac{1}{2} T_p S_p, \quad i, j \in \{1, 2, 3\}, \; p \in \{1, \ldots, 6\}.$$

2.4. Dynamic equation of an elastic body

2.4.1 Equations of vibrations

The problem is to find the motion of an elastic body attached in some way to the "surrounding" world, under the action of external forces, or in other words, subject to kinematic constraints and a mechanical loading.

From the static equations, the dynamic equations can be derived directly thanks to the d'Alembert principle, which supposes the second Newton's law as a static equilibrium of the body subjected to a virtual inertial force equal to minus the mass times acceleration (Lanczos, 2012). In the case of local equation, one must consider the density ρ, so the equations are:

$$-\rho\ddot{w}_i + T_{ij,j} + p_i = 0 \quad \text{in} \quad \Omega, \tag{2.37a}$$

$$w_i = 0 \quad \text{on} \quad \Sigma_k, \tag{2.37b}$$

$$T_{ij}n_j = \underline{\tau}_i \quad \text{on} \quad \Sigma_l \tag{2.37c}$$

where the surface Σ of the body is separated into two subsurfaces:

1. Σ_k where the kinematic constraints are defined by the imposed displacement $\underline{w}_i$ (here $w_i = 0$ means that no displacements are allowed);
2. Σ_l where a mechanical load is applied.

This set of constraints forms the boundary conditions that uniquely define the solution, namely the space–time function for the displacements. Actually, the problem can be simplified using some properties that we can observe regularly.

We experience daily the effects of vibration through hearing. When listening to an instrument playing, we can characterize a note by its pitch, but also we recognize the instrument by its timbre. These are two consequences of the most important properties of the vibration of an elastic body. For instance, as a string of a piano is hit by a hammer, its movement is a function of space and time that results from the superposition of many simpler movements. Each of them consists in specific shapes (functions of space variable only) modulated by harmonic functions of time at specific frequencies. They are totally fixed by the geometry of the string, its material, and the way they are attached to the structure of the piano (as stated by the equations above). Among all these frequencies, the lowest one defines the pitch, and the superposition of all the frequencies along with the time evolution of the amplitude of the shapes defines the timbre. These amplitudes are directly caused by the location and the intensity of the impact applied by the hammer.

So in the following we proceed by looking for the special solution that defines the special shapes and frequencies, and then find a way to reconstruct the overall solution by combining the response of these shapes to a mechanical loading.

2.4.2 Free harmonic vibrations

The free harmonic vibrations suppose to study the displacement of the body that is disturbed from its reference configuration at $t = 0$, assuming that no mechanical load are applied, i.e., $\underline{\tau}_i = 0$ for $t > 0$.

The trick is to seek the solutions of the problem as the product of a function of space and a function of time, that is, $\boldsymbol{w}(x_i, t) = \boldsymbol{\psi}(x_i)g(t)$. This method, called *variable separation method* (Arfken et al., 2013), is a common method used in solving partial differential equations. Actually, the solution is obtained as an infinite sum of functions having this form.

Substituting this form in Eq. (2.37a), and rearranging the terms, the following identity holds for each component of ψ:

$$\frac{\ddot{g}}{g} = \frac{\psi_i}{c_{ijkl}(\psi_{k,l} + \psi_{l,k})}. \tag{2.38}$$

Since this relation is verified for all t and x_i, these ratios must be equal to a constant. This implies that $g(t)$ is a *harmonic* function verifying $\ddot{g}(t) = -\omega^2 g(t)$ where ω represents the angular frequency of the vibration. Hence, this property is exploited to cancel out $g(t)$ of Eqs. (2.37a)–(2.37c), and the problem is now rewritten as

$$-\rho\omega^2\psi_i + c_{ijkl}S_{kl,j} = 0 \quad \text{in} \quad \Omega, \tag{2.39a}$$

$$\psi_i = 0 \quad \text{on} \quad \Sigma_k, \tag{2.39b}$$

$$\left[c_{ijkl}\left(\psi_{k,l} + \psi_{l,k}\right)\right] n_j = 0 \quad \text{on} \quad \Sigma_l. \tag{2.39c}$$

This set of partial differential equations with their boundary conditions forms a Sturm–Liouville problem (Arfken et al., 2013, Meirovitch, 1967). It can be proven for this class of problem that the solutions consist of eigenvalues (ω_k) associated to eigenfunctions ($\boldsymbol{\psi}^{(k)}(\boldsymbol{x})$), which are called resonant frequency and mode shape in the frame of elastic vibration, respectively. Moreover, the mode shapes $\boldsymbol{\psi}^{(k)}$ form a basis on Ω for all functions compatible with the kinematic conditions. This will be discussed in Sect. 2.4.4. We first illustrate the eigenvalues problem by an example.

Example 2 (part 1) Finding the mode shapes and the resonant frequencies

We consider the problem of the vibrations of a thin rod, fixed on one side, and subjected to a force on its free end. The problem is sketched in Fig. 2.4 (A). The material is supposed to be aluminum, which is orthotropic. The various parameters of the problem are summed up in Table 2.2. Initially, the rod is at rest so the displacement and the velocity fields are null for $t < 0$. We study the response to a step force applied at $t = 0$ on the free end of the rod.

The first step is to find the expression of the mode shapes and the corresponding frequencies. Even in the case of this simple example, there is no exact analytical solution, but it can be simplified with some assumptions.

Since the rod is supposed to be thin, and no loads are applied on its lateral faces, it is a reasonable assumption to let $T_1 = T_2 = 0$. Since the displacement is caused by a force applied along direction e_3, it is further assumed that S_3 only depends on x_3 and t. The compliance matrix of an orthotropic material such as aluminum is in this case

$$\boldsymbol{c} = \begin{bmatrix} c_{11} & c_{12} & c_{13} & 0 & 0 & 0 \\ c_{12} & c_{11} & c_{13} & 0 & 0 & 0 \\ c_{13} & c_{13} & c_{33} & 0 & 0 & 0 \\ 0 & 0 & 0 & c_{44} & 0 & 0 \\ 0 & 0 & 0 & 0 & c_{44} & 0 \\ 0 & 0 & 0 & 0 & 0 & c_{66} \end{bmatrix}.$$

It follows from the assumption that

$$\begin{aligned} S_1 &= -\frac{c_{13}}{c_{11} + c_{12}} S_3, \\ S_2 &= -\frac{c_{13}}{c_{11} + c_{12}} S_3, \\ T_3 &= c_{33} - 2\frac{c_{13}^2}{c_{11} + c_{12}} S_3. \end{aligned}$$

By identification we get $Y = c_{33} - 2\frac{c_{13}^2}{c_{11}+c_{12}} = \underline{c}_{33}$. So the displacement field is

$$\boldsymbol{w}(\boldsymbol{x}, t) = \begin{bmatrix} w_1(\boldsymbol{x}, t) = -\frac{c_{13}}{c_{11}+c_{12}} x_2 w_{,3}(x_3, t) \\ w_2(\boldsymbol{x}, t) = -\frac{c_{13}}{c_{11}+c_{12}} x_1 w_{,3}(x_3, t) \\ w_3(x_3, t) \end{bmatrix}. \tag{2.40}$$

The reader may verify that this displacement field verifies $S_4 = S_5 = S_6 = 0$ in static phase, but this is not the case in dynamic phase. The effects are, however, of a second order because the rod is thin and the strains are small. So in the remainder of this example, vibrations along directions e_1 and e_2 are disregarded.

We now apply Eqs. (2.37a)–(2.39c) keeping only the equation of motion along the e_3 component:

$$\begin{aligned} \rho \ddot{w}_3 - \underline{c}_{33} w_{3,33} &= 0, \\ w_3(0, t) &= 0, \\ \underline{c}_{33} w_{3,3}(L, t) &= \frac{F}{S}. \end{aligned}$$

As discussed, the solution is written as $w_3(x_3, t) = \psi_3(x_3)g(t)$, and it follows from the separation of variables that $\psi_3(x_3) = A\cos\beta x + B\sin\beta x$ and $g(t) = \sin(\omega t + \vartheta)$ with the following relationship:

$$\omega = \sqrt{\frac{c_{33}}{\rho}}\beta.$$

The initial phase ϑ at $t = 0$ is determined by the initial values of displacement and velocity fields, while the constants A and B are fixed by the spatial boundary conditions.

The modal analysis supposes the cancellation of the external force because no energy is supplied to the rod, and the boundary conditions are therefore:

$$A = 0,$$
$$\cos\beta L = 0,$$

where $g(t)$ is eliminated because the boundary conditions hold for all instants $t > 0$. This indicates that because the rod is finite, β must assume some specific values, which are denoted by β_n,

$$\beta_n = (2n-1)\frac{\pi}{2}\frac{1}{L}, \quad n \in \mathbb{N}^*.$$

A mode shape is associated to each of these values, so

$$\psi_3^{(n)}(x_3) = B_n \sin\left[(2n-1)\frac{\pi}{2}\frac{x_3}{L}\right],$$

and the resonant frequency associated to this solution is

$$\omega_n = \sqrt{\frac{c_{33}}{\rho}}\beta_n.$$

In Fig. 2.4 (B), the five first mode shapes are represented. With the parameters of Table 2.2, one can find the resonant frequencies, which are summed up in Table 2.1 for the five first modes.

2.4.3 Orthogonality of mode shapes

To examine the properties of these eigenfunctions, denoted by $\boldsymbol{\psi}^{(n)}$, which is the solution associated to the eigenvalue ω_n, form the functional (Geradin and Rixen, 2015)

$$\int_\Omega (-\rho\omega_n^2\psi_i^{(n)} + c_{ijkl}S_{kl,j}^{(n)})\psi_i^{(p)}\, d\Omega = -\int_\Omega \rho\omega_n^2\psi_i^{(n)}\psi_i^{(p)} + c_{ijkl}S_{kl}^{(n)}S_{ij}^{(p)}\, d\Omega. \quad (2.41)$$

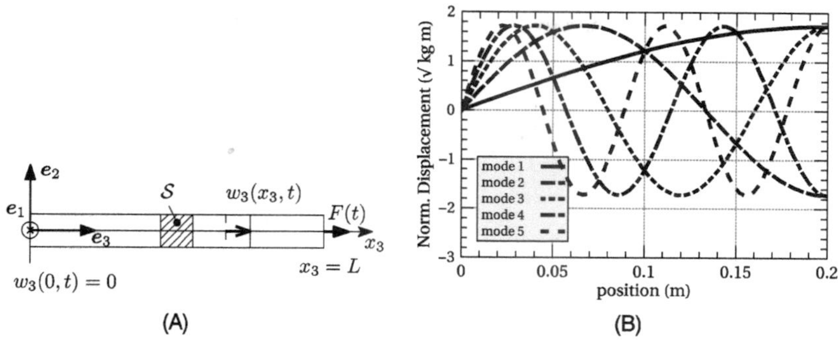

Figure 2.4 Schematic of the fixed-free thin rod and the longitudinal displacement (A) and the five first modes for longitudinal vibrations (B).

Table 2.1 Eigenvalues and resonant frequencies for the five first modes of the rod.

n	β_n (m^{-1})	f_n (Hz)
1	5π	1433
2	15π	4299
3	25π	7165
4	35π	9531
5	45π	11987

Table 2.2 Parameters used in the thin rod vibration example.

Density	$\rho = 2100$ kg/m^3
Young modulus	$Y = 69$ GPa
Modal damping	$\xi = 0.025$
Length	$L = 0.1$ m
Section	$S = 1 \cdot 10^{-4}$ m^2
Force	$F = 10$ N

The right-hand side of the equation is obtained by following the same arguments as in the previous section (using the divergence theorem and symmetry of the strains), and is not detailed here for brevity. Applying the same functional, where the role of the eigenfunctions are switched, leads to

$$\int_{\Omega} (-\rho\omega_p^2 \psi_i^{(p)} + c_{ijkl} S_{kl,j}^{(p)}) \psi_i^{(n)} = -\int_{\Omega} \rho\omega_n^2 \psi_i^{(p)} \psi_i^{(n)} + c_{ijkl} S_{kl}^{(p)} S_{ij}^{(n)} \, d\Omega. \quad (2.42)$$

Now, since $c_{ijkl} = c_{klij}$, we have the following identity: $c_{ijkl}S_{kl}^{(p)}S_{ij}^{(n)} = c_{klij}S_{ij}^{(p)}S_{kl}^{(n)} = c_{ijkl}S_{kl}^{(n)}S_{ij}^{(p)}$; therefore, taking the difference of Eqs. (2.41) and (2.42) leads to

$$(\omega_n^2 - \omega_p^2)\int_\Omega \rho\psi_i^{(n)}\psi_i^{(p)}d\Omega = 0. \tag{2.43}$$

Thus, as $\omega_n \neq \omega_p$, the previous result implies necessarily that for non-trivial solutions ($\boldsymbol{\psi}^{(n)} \neq 0$ and $\boldsymbol{\psi}^{(p)} \neq 0$)

$$\int_\Omega \rho\psi_i^{(n)}\psi_i^{(p)}d\Omega = 0 \quad \text{if} \quad n \neq p. \tag{2.44}$$

Since the functions $\boldsymbol{\psi}^{(n)}$ are eigenfunctions, they are defined up to a multiplicative constant (Arfken et al., 2013), which can be set to verify a given criterion. For now, the following choice is used:

$$\int_\Omega \rho\psi_i^{(n)}\psi_i^{(p)}d\Omega = 1. \tag{2.45}$$

This normalization simplifies result (2.44) to

$$\int_\Omega \rho\psi_i^{(n)}\psi_i^{(p)}d\Omega = \delta_{np}. \tag{2.46}$$

The normalized eigenfunctions $\boldsymbol{\psi}^{(n)}$ form an infinite set of *orthonormal* functions with respect to the scalar product defined by the previous equation. It is actually a complete basis for the space of functions verifying the boundary conditions (2.39b). Therefore, the fundamental *modal decomposition* property (or expansion theorem) holds for any admissible function $\boldsymbol{w}$ (Meirovitch, 1967):

$$\boldsymbol{w}(x, t) = \eta_n(t)\boldsymbol{\psi}^{(n)}(x_j) \tag{2.47}$$

where the $\eta_n(t)$ can be obtained by

$$\eta_n(t) = \int_\Omega \rho\psi_i^{(n)}w_i(x, t)d\Omega. \tag{2.48}$$

Finally, we conclude this discussion by noticing that normalization and orthogonality induce the following expression for the elastic energy:

$$2\mathcal{V}_e = \int_\Omega c_{ijkl}S_{kl}^{(n)}S_{ij}^{(n)}d\Omega, \tag{2.49}$$

which gives the property

$$\int_{\Omega} c_{ijkl} S_{kl}^{(p)} S_{ij}^{(n)} d\Omega = \int_{\Omega} T_{ij}^{(p)} S_{ij}^{(n)} d\Omega = \begin{cases} 0 & \text{if } n \neq p, \\ \omega_n^2 & \text{if } n = p. \end{cases} \tag{2.50}$$

Example 2 (part 2) Orthogonality property

We resume the previous example and verify the orthogonality properties. To do so, we first normalize the mode shapes to define B_n. In this simple example, B_n does not depend on n, that is, the modal mass is equal for all modes. Indeed, by applying the normalization with respect to the vibrating mass yields for B_n

$$\int_0^L B_n^2 \sin(\beta_n x_3)^2 \rho d\Omega = 1 \Leftrightarrow B_n = \sqrt{\frac{2}{\rho S L}} = \sqrt{\frac{2}{m}},$$

with $m = \rho S L$ being the mass of the rod. Consider now two mode shapes characterized by β_n and β_p, and write their projections as

$$\int \rho \psi^{(n)} \psi^{(p)} S dx_3 = \begin{cases} \rho S \frac{2}{m} \frac{L}{2} & = 1 \quad \text{if } \beta_n = \beta_p, \\ \rho S \frac{\beta_p \sin \beta_p L \cos \beta_n L - \beta_n \cos \beta_p L \sin \beta_n L}{\beta_p L^2 - \beta_n L^2} & = 0 \quad \text{if } \beta_n \neq \beta_p. \end{cases}$$

The first equality is true because of the normalization, while the second holds because in this example the boundary condition at $x_3 = L$ requires that $\cos \beta_n L = 0$ and $\cos \beta_p L = 0$. One can also verify the property (2.50):

$$\int \underline{c}_{33} \psi_{,3}^{(n)} \psi_{,3}^{(p)} S dx_3 =$$

$$\begin{cases} \underline{c}_{33} S \frac{2}{m} \frac{\beta_n^2 L}{2} = \frac{\underline{c}_{33}}{\rho} \beta_n^2 & = \omega_n^2 \quad \text{if } \beta_p = \beta_n, \\ \underline{c}_{33} S \frac{\beta_p \beta_n (\beta_p \sin \beta_p L \cos \beta_n L - \beta_n \cos \beta_p L \sin \beta_n L)}{\beta_p L^2 - \beta_n L^2} & = 0 \quad \text{if } \beta_m \neq \beta_n. \end{cases} \tag{2.51}$$

2.4.4 Forced vibrations

In order to study the response of an elastic body under loading, we apply the same procedure to Eq. (2.37a), that is, the dynamic forces field is projected on a given *normalized* mode shape giving

$$\int_{\Omega} \left(-\rho \ddot{w}_i + T_{ij,j} + p_i\right) \psi_i^{(n)} d\Omega = 0. \tag{2.52}$$

Integrating by parts and using the divergence theorem, one gets

$$\int_{\Omega} \left(-\rho \ddot{w}_i + T_{ij,j} + p_i\right) \psi_i^{(n)} d\Omega =$$
$$= \int_{\Sigma_l} \left(T_{ij} n_j\right) \psi_i^{(n)} d\Sigma + \int_{\Omega} p_i \psi_i^{(n)} d\Omega - \int_{\Omega} \rho \ddot{w}_i \psi_i^{(n)} + c_{ijkl} S_{kl} S_{ij}^{(n)} d\Omega = 0. \tag{2.53}$$

Using the decomposition property (2.48), the displacement field $\boldsymbol{w}$ and its corresponding strain are expressed as a linear combination of the mode shapes $\boldsymbol{\psi}^{(p)}$, hence

$$\boldsymbol{w}(\boldsymbol{x}, t) = \eta_p(t) \boldsymbol{\psi}^{(p)}(\boldsymbol{x}); \quad \ddot{\boldsymbol{w}}(\boldsymbol{x}, t) = \ddot{\eta}_p(t) \boldsymbol{\psi}^{(p)}(\boldsymbol{x});$$
$$S_{ij}(\boldsymbol{x}, t) = \eta_p(t) \frac{\psi_{i,j}^{(p)}(\boldsymbol{x}) + \psi_{j,i}^{(p)}(\boldsymbol{x})}{2} = \eta_p(t) S_{ij}^{(p)}(\boldsymbol{x}). \tag{2.54}$$

Substituting the latter expressions into Eq. (2.53) gives

$$\int_{\Omega} \left(-\rho \ddot{w}_i + T_{ij,j} + p_i\right) \psi_i^{(n)} d\Omega = \int_{\Sigma_l} \underline{\tau}_i \psi_i^{(n)} d\Sigma +$$
$$+ \int_{\Omega} p_i \psi_i^{(n)} d\Omega - \int_{\Omega} \rho \left(\ddot{\eta}_p(t) \psi_i^{(p)}\right) \psi_i^{(n)} + c_{ijkl} \left(\eta_p(t) S_{kl}^{(p)}\right) S_{ij}^{(n)} d\Omega = 0, \tag{2.55}$$

which is simplified using the orthogonality properties in the case of normalized mode shapes (2.46) and (2.50) into

$$\ddot{\eta}_n(t) + \omega_n^2 \eta_n(t) = \alpha_n(t), \tag{2.56}$$

where α_n is the modal "force", namely

$$\alpha_n = \int_{\Sigma_l} \underline{\boldsymbol{\tau}} \boldsymbol{\psi}^{(n)} d\Sigma + \int_{\Omega} \boldsymbol{p} \boldsymbol{\psi}^{(n)} d\Omega. \tag{2.57}$$

Thanks to this procedure, the problem now depends only on time. It remains to find the modal coordinates $\eta_n(t)$ as solutions of infinitely many elementary second-order oscillators excited by the modal force $\alpha_n(t)$ (Meirovitch, 1967). The solution can then be reconstructed using Eq. (2.47). Additionally, Eq. (2.57) implies that the ability of a given force distribution to excite a given mode depends on the mode shape.

Example 2 (part 3) Modal force and spatial filtering

Consider, for instance, the case where only a concentrated force is applied at some position $x_p = \frac{2}{3}L$ in the body, that is, $p(x, t) = P(t)\delta(x - x_p)$. Applying the definition results in $\alpha_n(t) = P(t)\eta_n(x_p)$. Applying this result to the two first modes, one gets $\alpha_1 = \sqrt{\frac{2}{m}}\frac{\sqrt{3}}{2}P$ and $\alpha_3 = 0$, thus the force cannot excite the second mode. Considering a physical interpretation of Eq. (2.57), it represents (ignoring a multiplication factor introduced by the normalization) the work produced by the external forces considering each mode separately. Here the force was applied on a node of the second mode, hence no energy can be transferred to it applying the load at this location. Likewise one can deduce that a force applied on the free end of the rod will excite all the modes (Preumont, 2011).

In addition to this spatial requirement for the location of forces, the time evolution is critical. To clarify this point, consider the Laplace transform of Eq. (2.56). Letting $H_n(s)$ and $A_n(s)$ denote the Laplace transforms of $\eta_n(t)$ and $\alpha_n(t)$, respectively, one gets

$$s^2 H_n(s) - s\,\dot{\eta}_n(0) - \eta_n(0) + \omega_n^2 H_n(s) = A_n(s). \tag{2.58}$$

The frequency response is thus the contribution of each mode, which will depend both on the modal forces and the initial conditions; therefore, according to the modal decomposition (2.47), one gets

$$\boldsymbol{w}(\boldsymbol{x}, s) = \left(\frac{A_n(s)}{s^2 + \omega_n^2} + \frac{s\,\dot{\eta}_n(0) - \eta_n(0)}{s^2 + \omega_n^2}\right)\boldsymbol{\psi}^{(n)}(\boldsymbol{x}). \tag{2.59}$$

In order to define the initial conditions, the initial shape $\boldsymbol{w}(\boldsymbol{x}, 0)$ and the initial speed field $\dot{\boldsymbol{w}}(\boldsymbol{x}, 0)$ must be projected on the modes according to Eq. (2.48). Without loss of generality, let the initial conditions be null. The total response is thus the superposition of the individual responses of second order low pass filters with cut-off frequencies ω_n. Therefore, the effectiveness of a given A_n to excite the nth mode is directly determined by its frequency spectrum.

Indeed, considering a sinusoidal excitation at one of the resonant angular frequency ω_n, according Eq. (2.59), the nth mode amplitude will grow indefinitely and therefore dominate all the others. This unrealistic infinite vibration is, of course, limited by the losses, which have not been considered so far. For a weakly damped material, the previous equation is rectified as follows (assuming $\dot{\eta}_n(0) = \eta_n(0)=0$ without loss of generality) to include the modal damping ξ_n:

$$\boldsymbol{w}(x, s) = \left(\frac{A_n(s)}{s^2 + 2\xi_n\omega_n s + \omega_n^2}\right)\boldsymbol{\psi}^{(n)}(x). \tag{2.60}$$

Example 2 (part 4) Response to a step load applied at the end of the rod

Suppose now that the loading consists of a force applied at the end of the rod at $t = 0$. This will cause a transient period involving a vibration before reaching the steady state deformation. In the case at hand, the modal forces can simply be written as

$$\alpha_n(t) = F(t)\sqrt{\frac{2}{m}}\sin\left[(2n-1)\frac{\pi}{2}\right] = (-1)^{n+1}\sqrt{\frac{2}{m}}F(t),$$

and one can notice that all the modes can be excited by such a load as claimed at the end of the previous example. The projection procedure in the case of a "step" force gives

$$\boldsymbol{w}(x, s) = \left(\frac{(-1)^{n+1}}{s^2 + 2\xi_n\omega_n s + \omega_n^2}\right)\frac{2}{m}\frac{F}{s}\sin\left[(2n-1)\frac{\pi}{2}\frac{x_3}{L}\right].$$

The frequency response at location $x_3 = L$ for a constant modal damping $\xi = 0.015$ (this value is chosen based on the quality factors usually measured for aluminum which are in a range of 30 to 70) is presented in Fig. 2.5 (A). The gray curve represents the frequency response to the force of the first mode ($n = 1$), while the black curves show the frequency response considering the five first modes. The modes clearly appear as resonance peaks, but also depend on the location. Considering now the location $x_3 = \frac{L}{5}$, the peak at frequency $f = 7165$ Hz disappears since it is a node of the third mode (cf. Fig. 2.5 (B)).

In practice, Eq. (2.60) must be truncated; however, in general, the convergence toward the solution can require a large number of modes. To address this issue, the solution can be sought using the quasistatic solution $\boldsymbol{w}^{(qs)}(\boldsymbol{x}, t)$ defined by the equations:

$$c_{ijkl}S^{(qs)}_{kl,j} + p_i = 0 \quad \text{in} \quad \Omega, \tag{2.61a}$$

$$w^{(qs)}_i = 0 \quad \text{on} \quad \Sigma_k, \tag{2.61b}$$

$$T^{(qs)}_{ij}n_j = \underline{\tau}_i \quad \text{on} \quad \Sigma_l. \tag{2.61c}$$

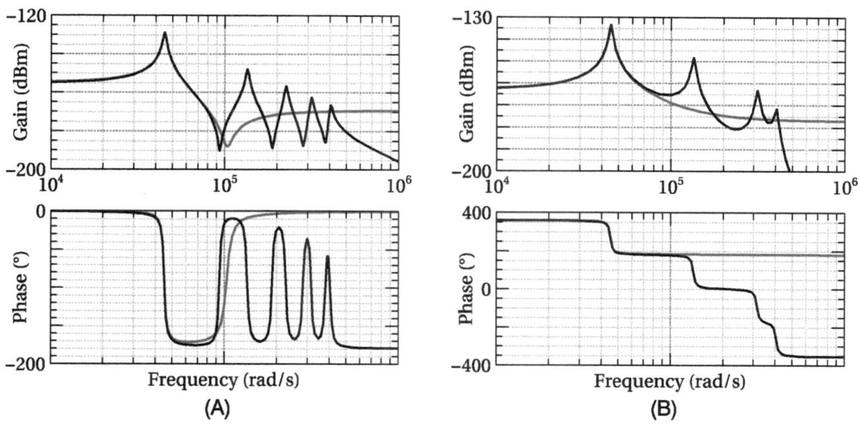

Figure 2.5 Frequency response of the displacement for a step of 10 N applied at $t = 0$ at position $x_3 = L$ (see (A)), and at $x_3 = \frac{L}{5}$ (see (B)) the second node of mode 3, the peak corresponding to its frequency, disappears.

It represents the displacement under mechanical loading, neglecting the inertia forces, which verifies the same kinematic constraints of the original problem. Therefore, the complete solution is of the form

$$\boldsymbol{w}(\boldsymbol{x}, t) = \boldsymbol{w}^{(qs)}(\boldsymbol{x}, t) + \boldsymbol{\psi}^{(n)}(\boldsymbol{x})\upsilon_n(t) \tag{2.62}$$

where $\upsilon_n(t)$ are the new modal coordinates. They are determined by substituting the solution (2.62) into Eq. (2.52), which after some algebra gives

$$\ddot{\upsilon}_n(t) + \omega_n^2 \upsilon_n(t) = \frac{d^2}{dt^2}\tilde{\alpha}_n(t), \tag{2.63}$$

where

$$\tilde{\alpha}_n(t) = -\int_{\Omega} \rho \boldsymbol{w}^{(qs)}(\boldsymbol{x}, t) \cdot \boldsymbol{\psi}^{(n)}(\boldsymbol{x}) d\Omega. \tag{2.64}$$

To compare the two methods, the solution is written using the Laplace transform. For the sake of simplicity and without loss of generality, the body is supposed to be initially at rest (i.e., $\upsilon(0) = \dot{\upsilon}(0) = 0$), so

$$\begin{aligned} \boldsymbol{W}(\boldsymbol{x}, s) &= \boldsymbol{W}^{(qs)}(\boldsymbol{x}, s) + \frac{s^2 \tilde{A}_n(s)}{s^2 + \omega_n^2} \boldsymbol{\psi}^{(n)}(\boldsymbol{x}) = \\ &= \boldsymbol{W}^{(qs)}(\boldsymbol{x}, s) - \frac{s^2}{s^2 + \omega_n^2} \left(\int_{\Omega} \rho \boldsymbol{W}^{(qs)}(\boldsymbol{x}, s) \cdot \boldsymbol{\psi}^{(n)}(\boldsymbol{x}) d\Omega \right) \boldsymbol{\psi}^{(n)}(\boldsymbol{x}), \end{aligned} \tag{2.65}$$

with $\tilde{A}_n(s)$ being the Laplace transform of $\tilde{\alpha}_n(t)$.

Compared to the solution (2.59), the dynamically compensated quasistatic response is the superposition of the static response and high pass second-order filters; therefore, the requirement for higher order modes to describe correctly the low frequencies' contribution is not needed any more (Geradin and Rixen, 2015).

Noticing that the quasi static displacement can also be decomposed on the modal basis, i.e.,

$$\boldsymbol{W}^{(qs)}(\boldsymbol{x}, s) = \left(\int_{\Omega} \rho \boldsymbol{W}^{(qs)}(\boldsymbol{x}, s) \cdot \boldsymbol{\psi}^{(n)}(\boldsymbol{x}) d\Omega \right) \boldsymbol{\psi}^{(n)}(\boldsymbol{x}),$$

one can rewrite Eq. (2.65) as

$$\boldsymbol{W}(\boldsymbol{x}, s) = \frac{\omega_n^2 \int_{\Omega} \rho \boldsymbol{W}^{(qs)}(\boldsymbol{x}, s) \cdot \boldsymbol{\psi}^{(n)}(\boldsymbol{x}) d\Omega}{s^2 + \omega_n^2} \boldsymbol{\psi}^{(n)}(\boldsymbol{x}).$$

Comparing this result to Eq. (2.59), one gets the relationship between $\tilde{\alpha}_n$ and α_n:

$$-\omega_n^2 \tilde{\alpha}_n(t) = \alpha_n(t). \tag{2.66}$$

Example 2 (final part) Using the quasi-static solution

For the thin rod, the quasistatic solution for displacement is linear since its second derivative is null. Taking into account the boundary conditions, one gets

$$w_3^{(qs)}(x_3, t) = \frac{F(t)}{c_{33} S} x_3.$$

According to Eq. (2.64),

$$\begin{aligned} \tilde{\alpha}_n(t) &= -\rho S \int \frac{F(t)}{c_{33} S} x_3 \sqrt{\frac{2}{m}} \sin\left[(2n-1)\frac{\pi}{2}\frac{x_3}{L}\right] dx_3 \\ &= -\frac{\rho}{c_{33}} \sqrt{\frac{2}{m}} F(t) \frac{(-1)^{n+1}}{\beta_n^2}. \end{aligned}$$

Owing to the relationship $\omega_n^2 = \frac{\rho}{c_{33}} \beta_n^2$ found in the first part of the example, this simplifies to

$$\tilde{\alpha}_n(t) = (-1)^n \frac{1}{\omega_n^2} \sqrt{\frac{2}{m}} F(t).$$

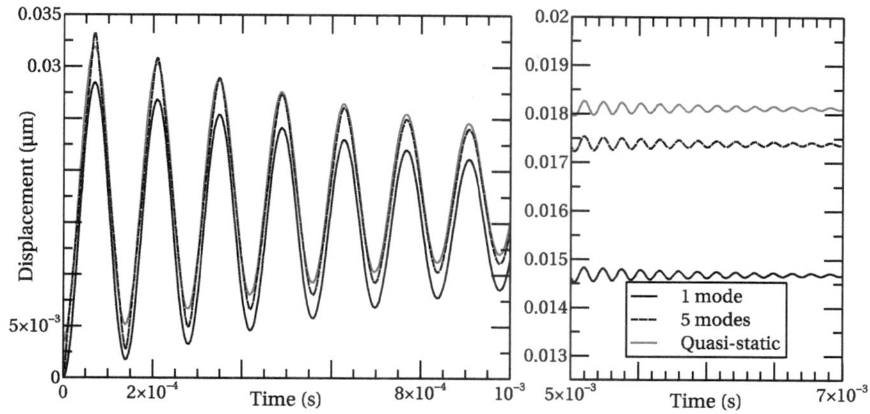

Figure 2.6 Comparison of the time response of the direct projection method for one and five modes vs the quasistatic approach with one mode: taking only one mode into account, in the direct projection method underestimates the transient amplitude and wrongly predicts the steady state. Five modes are required to achieve less than 5% error. By comparison the transient of the quasistatic method model is less accurate to the actual one, but it is, by construction, exact at steady state.

The reader can easily verify from part 3 of the example that Eq. (2.66) indeed holds.

Finally, an alternative expression for the step loading response is

$$w_3(x_3, s) = \frac{F}{s}\left(\frac{x_3}{\underset{\sim}{c}_{33}S} + \sum_{n=1}^{\infty}\frac{1}{\omega_n^2}\frac{s^2}{s^2+2\xi_n\omega_n s+\omega_n^2}\frac{2(-1)^n}{m}\sin\left[(2n-1)\frac{\pi}{2}\frac{x_3}{L}\right]\right).$$

In Fig. 2.6, the transient responses for the modal truncation using one mode (solid curves) and five modes (dashed curves) are compared to the quasistatic decomposition using one mode (grey curve). The quasistatic approach predicts the steady state exactly by construction. Using the first approach requires 5 modes to achieve 5% accuracy. On the other hand, the transient waveform is not so well predicted for the quasistatic solution because only one mode is used. This is visible by examining the first overshoot which is less sharp, but, as the high frequencies' contributions of the modal decomposition rapidly decay compared to the first mode, the error quickly vanishes.

2.5. Dynamic of piezoelectric bodies

Piezoelectric materials combine electric and elastic coupling. In this section, the main electrostatic properties needed to describe this phenomenon will be recalled, then the constitutive laws describing the coupling between the stress, strain, electric field and electric induction field will be presented. Finally, an extension of the modal analysis to piezoelectric media is explained.

2.5.1 Basic of electrostatic

2.5.1.1 Electric field and electric potential

Two electric charges q_1, q_2 located at points A and B in the void are subjected to electrostatic forces which are along the straight line joining them, and are attractive if the charges have the same sign, repulsive otherwise. The Coulomb law gives the force as a function of the distance separating them. Let $\boldsymbol{r}_{1\to 2}=\boldsymbol{AB}$ and $r=\|\boldsymbol{AB}\|$. Then the force exerted by q_1 on q_2 is

$$\boldsymbol{F}_{1\to 2}=\frac{1}{4\pi\epsilon_0}\frac{q_1\times q_2}{r^2}\frac{\boldsymbol{r}_{1\to 2}}{r}, \tag{2.67}$$

ϵ_0 here is the permittivity of the void. For more general charge distribution, it is convenient to introduce the electric field $\boldsymbol{E}$ such that the electrostatic force on a particular charge q_i is

$$\boldsymbol{F}_i=q_i\boldsymbol{E}. \tag{2.68}$$

The electrostatic force is conservative. Hence, the electric vector field $\boldsymbol{E}$ resulting from a given distribution of charge in space is derived from a scalar electric potential ϕ:

$$\boldsymbol{E}=-\nabla\phi \Leftrightarrow E_i=-\phi_{,i}, \quad i\in\{1,2,3\}. \tag{2.69}$$

In practice, it is convenient to define the charge distribution using a charge density ϱ. In the void, the electric field flux is conservative, that is, the total electric flux through a regular closed surface is proportional to the total electric charge it encloses. This is the Gauss theorem (Owen, 2013):

$$\int_{\Sigma}\boldsymbol{E}\boldsymbol{n}d\Sigma=\frac{1}{\epsilon_0}\int_{\Omega}\varrho d\Omega, \tag{2.70}$$

where $\boldsymbol{n}$ is the outward normal defined on Σ, and Ω is the volume enclosed within Σ. This can be reformulated using the divergence theorem to give the local Gauss theorem:

$$\text{div}\boldsymbol{E} = \nabla \cdot \boldsymbol{E} = \frac{\varrho}{\epsilon_0}. \quad (2.71)$$

Given property (2.69), this leads to the expression relating the electric potential ϕ to the charge density, namely

$$\nabla^2 \phi + \frac{\varrho}{\epsilon_0} = 0, \quad (2.72)$$

which is the Poisson's equation.

2.5.1.2 Interaction of the electric field with conductive and dielectric media

The piezoelectric effect is an interaction inside matter. From the electrostatic point of view, two important classes of media are the conductor and dielectric (Pramanik, 2008), which are both involved in realizing piezoelectric devices. The conductors are used for the electrodes that define the electric equipotential conditions on the boundaries of the piezoelectric ceramic (sometimes inside in the case of multilayer technologies), which is dielectric.

In conductors, electric charges are free to move under the effect of the external electric field. As a consequence, once a conductor is at electric equilibrium, the charges are distributed in such a way that they cancel the external electric field within the conductor. Denoting the external electric field $\boldsymbol{E}_{ext}$, and the electric field resulting from the electric charge distribution of the conductor $\boldsymbol{E}_{int}$, the total field is null, that is, $\boldsymbol{E} = \boldsymbol{E}_{ext} + \boldsymbol{E}_{int} = 0$. Therefore the electric potential in the conductor is constant since, according to Eq. (2.69), $\nabla\phi = 0$. Moreover, using this in the local Gauss theorem (2.71) yields $\frac{\varrho}{\epsilon_0} = 0$. Hence there are no charges inside the conductor, which means that all the charges are located on the surface of the conductor. Finally, since the conductor is at a constant electric potential, its surface defines an equipotential, and consequently the electric field is normal to the surface of the conductor.

In contrast, the dielectric materials are nonconductive, in other words, there are no free electric charges in an ideal dielectric body. However, matter at a molecular or atomic scale includes positive and negative charges. When subjected to an electric field, they are subjected to forces and, due to their opposite charges, tend to be separated. Therefore, locally, the atom or molecule charges are no longer coincident, rather they have a dipolar nature. One can therefore define a dipolar moment $\boldsymbol{m}$: for two opposite charges located at A and B, $\boldsymbol{m} = q\boldsymbol{AB}$. In this representation, the distance separating A and B is small compared to the position P where the potential is calculated, namely $AB \ll MP$, where M is the barycenter of the charges.

The electric dipole potential is given by $\phi(P) = \frac{1}{4\pi\epsilon_0}\left(\frac{1}{r_A} - \frac{1}{r_B}\right)$, with $\|AP\| = r_A$ and $\|BP\| = r_B$ being the distances from A and B, respectively. In a first order approximation, the potential can be rewritten as $V(r) = \frac{1}{4\pi\epsilon_0}\nabla(r^{-1}) \cdot \boldsymbol{m}$, where $r = \|MP\|$.

In the framework of continuum mechanics, the polarization vector $\boldsymbol{P}$ is introduced to represent a dipole density. The dipolar moment is then given by $\boldsymbol{m} = \boldsymbol{P}d\Omega$, and the potential can be rewritten as $\phi(r) = \frac{1}{4\pi\epsilon_0}\nabla(r^{-1}) \cdot \boldsymbol{P}d\Omega$. Considering a dielectric volume Ω, the total potential can be calculated by summing the potential due to the polarization in each elementary volume. Integrating over the volume and using the divergence theorem finally gives

$$V(r) = \frac{1}{4\pi\epsilon_0}\oint_{\Sigma}\frac{\boldsymbol{P}\cdot\boldsymbol{n}}{t}d\Sigma + \frac{1}{4\pi\epsilon_0}\int_{\Omega}\frac{-\nabla\cdot\boldsymbol{P}}{r}d\Omega. \tag{2.73}$$

To make this last equation clearer, the equivalent surface density of charge on the boundary $\varrho = \boldsymbol{P}\cdot\boldsymbol{n}$ and the volume charge density $\underline{\varrho} = -\nabla\cdot\boldsymbol{P}$ are introduced. With this in mind, it appears that when subjected to an external electric field, the dielectric behaves as a body having an induced volume and surface charge density. Therefore, it follows from the local Gauss theorem (2.71) and the fact that no free charges exist in a dielectric that locally the electric field is

$$\nabla\cdot\boldsymbol{E} = -\frac{\nabla\cdot\boldsymbol{P}}{\epsilon_0}, \tag{2.74}$$

so that it is no longer flux-conservative due to the microscopic polarization. To measure the combined effects of the external field and internal polarization, the electric induction vector $\boldsymbol{D}$ is defined as follows:

$$\boldsymbol{D} = \epsilon_0\boldsymbol{E} + \boldsymbol{P}. \tag{2.75}$$

The behavior law of a given material is characterized using the susceptibility tensor $\boldsymbol{\chi}$ that relates the polarization $\boldsymbol{P}$ to the external field $\boldsymbol{E}$ as

$$P_i = \epsilon_0\chi_{ij}E_j.$$

Thus, the components of the electric induction are

$$D_i = \epsilon_0\left(1 + \chi_{ij}\right)E_j = \epsilon_{ij}E_j \tag{2.76}$$

where the permittivity tensor $\boldsymbol{\epsilon}$ is introduced, having components $\epsilon_{ij} = \epsilon_0 (1 + \chi_{ij})$. Combining (2.74) and the definition (2.75) yields the important property

$$\nabla \boldsymbol{D} = 0. \tag{2.77}$$

To sum up, in a dielectric:

- The electric field $\boldsymbol{E}$ derives from a potential, but its flux is no longer conservative;
- The electric displacement field $\boldsymbol{D}$ does not depend on a potential (due to the polarization $\boldsymbol{P}$); however, its flux is conservative.

Finally, to conclude this short review of the main properties of electrostatics useful in the understanding of piezoelectricity, we recall the crossing conditions between two media sharing a boundary. At the boundary separating the two media, the following continuity conditions hold:

- The tangential components of the electric fields are equal;
- The normal components of the electric displacements are equal.

Here normal and tangential refer to the local frame defined along the boundary. Considering now one dielectric media sharing a boundary with a conductor having the surface charge density ϱ, these requirements result in the following crossing conditions:

$$E_t = 0, \tag{2.78}$$

$$D_n = -\varrho. \tag{2.79}$$

2.5.2 Piezoelectricity constitutive equations

The piezoelectricity constitutive laws that relate the mechanical tensors $\boldsymbol{T}$, $\boldsymbol{S}$ to the electric vectors $\boldsymbol{E}$ and $\boldsymbol{D}$ were discovered by Pierre and Jacques Curie. They studied the correlation between crystallographic structures of pyroelectric materials to explain its origins. Following the ideas devised by Forbes and Thomson, they tried to relate it to the polarization of the crystals which would explain why charges appear only on some of the faces.

During this work in 1880, they recognized that the variation of dimensions due to dilatation along directions coinciding with some of the symmetry axes of such crystals was the origin of the produced charges. This led to a series of communications regarding the production of charges proportional to pressure variation, their polarity according to compression or dilatation and the importance of the direction of the strain with regard to the symmetry axis of the crystals to produce charges. They also observed that, on two opposite faces, the produced charges had opposite signs. In

1881, they published some results that linearly related the produced charge with the length variation of the sample.

Finally, in 1882, they confirmed the arguments of Lippmann who, considering charge and energy conservation, predicted that to verify the laws of thermodynamic, the phenomenon should be reversible. The Curie brothers then designed an ingenious set-up to measure the minute displacements of a quartz crystal and observed the so-called "inverse" effect.

The piezoelectricity constitutive laws assume a linear coupling between elasticity and electrostatic. They therefore described the material's behavior in the vicinity of a particular state defined by its initial polarization. Moreover, this coupling is a part of a more complex interaction which involves temperature as well, as the history of its discovery suggests. However, in the following, pyroelectricity is not taken into account.

Selecting the extensive quantities $\boldsymbol{S}$ and $\boldsymbol{D}$ as *independent* variables, the constitutive laws are given by (Ikeda, 1996)

$$T_{ij} = c^D_{ijkl} S_{kl} - h_{ijm} D_m, \tag{2.80a}$$

$$E_n = -h_n klS_{kl} + \beta^S_{nm} D_m, \tag{2.80b}$$

where c^D_{ijkl} are the rigidities at constant electric displacement field, β^S_{nm} are the inverse permittivities at constant strain, and h_{nkl} are the piezoelectric coefficients. The piezoelectric coupling is reversible, hence these relations derive from a potential. In this case, the total differential is given by

$$dF = TdS + EdD \Leftrightarrow F(S, D) = \frac{1}{2} c^D_{ijkl} S_{ij} S_{kl} - h_{nij} D_n S_{ij} + \frac{1}{2} \beta^S_{nm} D_n D_m. \tag{2.81}$$

Moreover, reversibility implies that $h_{nij} = h_{ijn}$. It is actually possible to write the constitutive laws differently using a Legendre transformation (Lanczos, 2012) in order to address a problem using the relevant independent variables. These laws, along with their associated potentials, are summed up in Table 2.3. Considering the energy potential given by (2.81), it appears that it consists of three contributions:

- the elastic energy $\mathcal{V}_M = \frac{1}{2} c^D_{ijkl} S_{ij} S_{kl}$,
- the electrostatic energy $\mathcal{V}_E = \frac{1}{2} \beta^S_{nm} D_n D_m$,
- the interaction energy $\mathcal{V}_I = -h_{nij} D_n S_{ij}$.

The first two energies are essentially stored energies that will therefore be returned to the environment that caused their modifications. Only the interaction energy is responsible for an electromechanical energy conversion. In order to do so, the interaction energy must be modified by acting on $\boldsymbol{S}$

Table 2.3 The different constitutive laws of piezoelectricity and their associated thermodynamic potentials.

Independent variables	Constitutive law	Potential differential
$\boldsymbol{S}, \boldsymbol{D}$	$\boldsymbol{T} = \boldsymbol{c}^D \boldsymbol{S} - \boldsymbol{h}\boldsymbol{D}$ $\boldsymbol{E} = -\boldsymbol{h}\boldsymbol{S} + \boldsymbol{\beta}^S \boldsymbol{D}$	$dF = \boldsymbol{T} d\boldsymbol{S} + \boldsymbol{E} d\boldsymbol{D}$
$\boldsymbol{T}, \boldsymbol{E}$	$\boldsymbol{S} = \boldsymbol{s}^E \boldsymbol{T} + \boldsymbol{d}\boldsymbol{E}$ $\boldsymbol{D} = \boldsymbol{d}\boldsymbol{T} + \boldsymbol{\epsilon}^T \boldsymbol{E}$	$dG = -\boldsymbol{S} d\boldsymbol{T} - \boldsymbol{D} d\boldsymbol{E}$
$\boldsymbol{T}, \boldsymbol{D}$	$\boldsymbol{S} = \boldsymbol{s}^D \boldsymbol{T} + \boldsymbol{g}\boldsymbol{D}$ $\boldsymbol{E} = -\boldsymbol{g}\boldsymbol{T} + \boldsymbol{\beta}^T \boldsymbol{D}$	$dG_1 = -\boldsymbol{S} d\boldsymbol{T} + \boldsymbol{E} d\boldsymbol{D}$
$\boldsymbol{S}, \boldsymbol{E}$	$\boldsymbol{T} = \boldsymbol{c}^E \boldsymbol{S} - \boldsymbol{e}\boldsymbol{E}$ $\boldsymbol{D} = \boldsymbol{e}\boldsymbol{S} + \boldsymbol{\epsilon}^S \boldsymbol{E}$	$dG_2 = \boldsymbol{T} d\boldsymbol{S} - \boldsymbol{D} d\boldsymbol{E}$

or $\boldsymbol{D}$, which implies that the simultaneously stored energy must be modified electrically and/or mechanically. It is thus impossible to realize a cycle that would totally cancel the stored energy without concomitantly canceling the interaction energy. *Consequently, the amount of energy converted can only be a fraction of the total energy put in the piezoelectric during a cycle.* The coupling factor is defined as the square root of the ratio between the converted and the stored energy. It is a merit figure for a given piezoelectric material that can be found on the manufacturer's datasheet. From this discussion, it can be deduced that the coupling factor is always less than unity.

2.5.3 Electromechanical dynamic equations

The dynamic problem of a piezoelectric body involves solving simultaneously the dynamic equation elasticity and the electrostatic problem, given that they are coupled through the piezoelectric laws (Tiersten, 2013).

Hence the equilibrium in a volume of body Ω implies that the following equations are simultaneously verified for $\boldsymbol{x} \in \Omega$:

$$\nabla \cdot \boldsymbol{T} + \boldsymbol{p} = \rho \ddot{\boldsymbol{w}}, \tag{2.82a}$$

$$r\nabla \cdot \boldsymbol{D} = 0. \tag{2.82b}$$

On the mechanical side, the kinematic constraints and the forces loading the body must be specified on the respective surfaces Σ_k and Σ_l:

$$\boldsymbol{T} \cdot \boldsymbol{n} = \underline{\boldsymbol{\tau}} \qquad \text{for } \boldsymbol{x} \in \Sigma_l, \tag{2.83a}$$

$$\boldsymbol{w} = 0 \qquad \text{for } \boldsymbol{x} \in \Sigma_k. \tag{2.83b}$$

On the electric side, some parts of the body do not have electrodes, and no free charges can appear. They will be denoted $\Sigma_{\bar{e}}$. The electrodes, on

Σ_e, define equipotentials, but also the total charge on them must be equal to the charges supplied by the generators connected to them. According to whether a given generator imposes a charge or voltage, the boundary conditions must be adapted. In the following, only the second case is examined as it is the most frequently encountered.

In this case the electrical boundary conditions are:

$$\phi = v \qquad \text{for } \in \Sigma_e, \tag{2.84a}$$

$$\boldsymbol{D} \cdot \boldsymbol{n} = 0 \qquad \text{for } \in \Sigma_{\bar{e}}. \tag{2.84b}$$

In addition to Eqs. (2.82a)–(2.84b), the relations linking the electric field to the electric potential (2.69) and the strain to the displacements (2.9) are also introduced.

For instance, suppose that the chosen independent variables to express the piezoelectricity constitutive laws are the electric field $\boldsymbol{E}$ and strain $\boldsymbol{S}$, then the dynamic equations are reduced to four equations with four unknowns, namely the displacement w_j, $j \in \{1, 2, 3\}$, and the electric potential ϕ. Omitting the volumic force for the sake of simplicity and without loss of generality, the equations are:

$$c_{ijkl}^{E} w_{k,lj} + e_{ijp}\phi_{,pj} = \rho \ddot{w}_i, \tag{2.85a}$$

$$e_{pij} w_{i,jp} - \epsilon_{qp}^{S} \phi_{,qp} = 0. \tag{2.85b}$$

Meanwhile, some of the boundary conditions can be rewritten to keep only the relevant variables as:

$$\left(c_{ijkl}^{E} w_{k,l} + e_{ijp}\phi_{,p}\right) n_j = \underline{\tau}_i \qquad \text{for } \boldsymbol{x} \in \Sigma_l, \tag{2.86a}$$

$$w_i = 0 \qquad \text{for } \boldsymbol{x} \in \Sigma_k, \tag{2.86b}$$

$$\phi = v \qquad \text{for } \boldsymbol{x} \in \Sigma_e, \tag{2.86c}$$

$$\left(e_{pij} w_{i,j} - \epsilon_{q}^{S} \phi_{,q}\right) n_p = 0 \qquad \text{for } \boldsymbol{x} \in \Sigma_{\bar{e}}. \tag{2.86d}$$

2.5.4 Free harmonic equation and orthogonality

The free harmonic vibration can be studied using the previous equations with $\underline{\tau}_i = 0$ and $v = 0$. To apply the separation of variables, it will be assumed that the solutions are now written as $\boldsymbol{w}(\boldsymbol{x}, t) = \boldsymbol{\psi}(\boldsymbol{x})g(t)$ with $\boldsymbol{\psi} = \left[\psi_1, \psi_2, \psi_3\right]^{\top}$ *and* $\phi(\boldsymbol{x}, t) = \varphi(\boldsymbol{x})g(t)$ where φ is the electric potential for free vibrations. As before, $g(t)$ is a harmonic function, so $\ddot{g}(t) = -\omega^2 g(t)$.

The equations obtained are again a Sturm–Liouville problem. Hence, infinitely many solutions, each one consisting of a resonant angular frequency ω_n and a generalized mode shape $\boldsymbol{\Psi}^{(n)} = \left[\psi_1^{(n)}, \psi_2^{(n)}, \psi_3^{(n)}, \varphi^{(n)}\right]^{\mathrm{T}}$, can be found (here n denotes a particular solution). The only difference is that the electric potential is now part of the mode shape. The equilibrium equations verified by the modes on Ω are:

$$c_{ijkl}^{E}\psi_{k,lj}^{(n)} + e_{ipj}\varphi_{,pj}^{(n)} = -\rho\omega_n^2\psi_i^{(n)}, \tag{2.87a}$$

$$e_{pij}\psi_{i,jp}^{(n)} - \epsilon_{qp}\varphi_{,qp}^{(n)} = 0, \tag{2.87b}$$

with

$$\left(c_{ijkl}^{E}\psi_{k,l}^{(n)} + e_{ipj}\varphi_{,p}^{(n)}\right) n_j = 0 \qquad \text{for } \boldsymbol{x} \in \Sigma_l, \tag{2.88a}$$

$$\psi_i^{(n)} = 0 \qquad \text{for } \boldsymbol{x} \in \Sigma_k, \tag{2.88b}$$

$$\varphi^{(n)} = 0 \qquad \text{for } \boldsymbol{x} \in \Sigma_e, \tag{2.88c}$$

$$\left(e_{pij}\psi_{i,j}^{(n)} - \epsilon_{qp}\varphi_{,q}^{(n)}\right) n_p = 0 \qquad \text{for } \boldsymbol{x} \in \Sigma_{\bar{e}}. \tag{2.88d}$$

Due to the similarities highlighted so far, one can expect that the orthogonality property also holds for the "piezoelectric modes". To prove this, consider the following functional applied to two modes n and p:

$$\mathcal{P}_{\psi^{(n)}}\left(\psi^{(p)}\right) = \int_{\Omega}\left(T_{ij,j}^{(p)} + \rho\omega_n^2\psi_i^{(p)}\right)\psi_i^{(n)} + D_{i,i}^{(p)}\varphi^{(n)}\, d\Omega = 0. \tag{2.89}$$

This functional is necessarily null since $\boldsymbol{\psi}_m^{(p)}$ verifies Eqs. (2.87a) and (2.87b). The volume integral of the scalar product $T_{ij,j}\psi_i^{(n)}$ is rearranged using the divergence theorem and stress tensor symmetry, while the volume integral of the product $D_{i,i}^{(p)}\varphi^{(n)}$ is transformed thanks to the identity

$$\nabla \cdot (a\mathbf{V}) = (\nabla a) \cdot \mathbf{V} + a(\nabla \cdot \mathbf{V})$$

where a is a scalar and $\boldsymbol{V}$ a vector. This leads to

$$\mathcal{P}_{\psi^{(n)}}\left(\psi^{(p)}\right) = \int_{\Sigma}\left(T_{ij}^{(p)}n_j\right)\psi_i^{(n)} + \left(D_i^{(p)}n_i\right)\varphi^{(n)}\, d\Sigma + \\ + \int_{\Omega}\rho\omega_p^2\psi_i^{(p)}\psi_i^{(n)} - T_{ij}^{(p)}S_{ij}^{(n)} - D_i^{(p)}\varphi_{,i}^{(n)}\, d\Omega. \tag{2.90}$$

Since the boundary conditions are null, the surface integral is dropped from this equation, and introducing the electric field (2.69) finally gives

$$\mathcal{P}_{\psi^{(n)}}\left(\psi^{(p)}\right)=\int_{\Omega}\rho\omega_p^2\psi_i^{(p)}\psi_i^{(n)}-T_{ij}^{(p)}S_{ij}^{(n)}+D_i^{(p)}E_i^{(n)}d\Omega=0. \tag{2.91}$$

Applying the same functional while swapping the two modes results in swapping the indexes n and p in the previous expression. Thus, subtracting both results gives

$$\mathcal{P}_{\psi^{(n)}}\left(\psi^{(p)}\right)-\mathcal{P}_{\psi^{(p)}}\left(\psi^{(n)}\right)=\int_{\Omega}\rho\left(\omega_p^2-\omega_n^2\right)\psi_i^{(n)}\psi_i^{(p)}-\left(T_{ij}^{(p)}S_{ij}^{(n)}-T_{ij}^{(n)}S_{ij}^{(p)}\right)$$
$$+\left(D_i^{(p)}E_i^{(n)}-D_i^{(n)}E_i^{(p)}\right)d\Omega=0. \tag{2.92}$$

Considering the $\boldsymbol{S}$, $\boldsymbol{E}$ piezoelectric law, one gets

$$\int_{\Omega}\rho\left(\omega_p^2-\omega_n^2\right)\psi_i^{(n)}\psi_i^{(p)}-\left(c_{ijkl}S_{kl}^{(p)}S_{ij}^{(n)}-c_{ijkl}S_{kl}^{(n)}S_{ij}^{(p)}\right)+$$
$$+\left(e_{kij}E_k^{(p)}S_{ij}^{(n)}-e_{ikl}E_i^{(n)}S_{kl}^{(p)}\right)+\left(e_{kij}E_k^{(n)}S_{ij}^{(p)}-e_{ikl}E_i^{(p)}S_{kl}^{(n)}\right)+$$
$$+\left(\epsilon_{ik}E_i^{(p)}E_k^{(n)}-\epsilon_{ik}E_i^{(n)}E_k^{(p)}\right)d\Omega=0. \tag{2.93}$$

However, using the symmetry of the tensors, and since the index notations are arbitrary, this equation reduces to

$$\mathcal{P}_{\psi^{(n)}}\left(\psi^{(p)}\right)-\mathcal{P}_{\psi^{(p)}}\left(\psi^{(n)}\right)=\left(\omega_p^2-\omega_n^2\right)\int_{\Omega}\rho\psi_i^{(p)}\psi_i^{(n)}d\Omega=0, \tag{2.94}$$

which can hold for $p\neq n$ only if

$$\int_{\Omega}\rho\psi_i^{(p)}\psi_i^{(n)}d\Omega=0, \tag{2.95}$$

proving the orthogonality of the displacement field for the piezoelectric vibration modes. Applying the normalization (2.45), property (2.50) can be extended to the piezoelectric medium as follows:

$$\int_{\Omega}T_{ij}^{(p)}S_{ij}^{(n)}-D_i^{(p)}E_i^{(n)}d\Omega=\begin{cases}\omega_p^2 & \text{if } p=n,\\ 0 & \text{if } p\neq n.\end{cases} \tag{2.96}$$

2.5.5 Forced vibrations

The projection method can be applied to the piezoelectric media dynamic solution, as in the case of the elastic media. The main difference is that now the eigenfunctions include both mechanical displacement and an electrical quantity. The solution is sought in the form of a superposition of the modes.

2.5.5.1 Imposed charges

For instance, if the electric potential is considered as a generalized coordinate, the solution is expressed as

$$\boldsymbol{W}(\boldsymbol{x},t)=\begin{bmatrix}\boldsymbol{w}(\boldsymbol{x},t)\\ \varphi(\boldsymbol{x},t)\end{bmatrix}=\begin{bmatrix}\boldsymbol{\psi}^{(n)}(\boldsymbol{x})\\ \varphi^{(n)}(\boldsymbol{x})\end{bmatrix}\eta_n(t)=\boldsymbol{\Psi}^{(n)}(\boldsymbol{x})\eta_n(t). \tag{2.97}$$

In order to find the dynamic differential equation governing the modal coordinates η_n, the projection is performed as previously:

$$\mathcal{P}_{\Psi^{(n)}}(\boldsymbol{W})=\int_\Omega \{-\rho\ddot{w}_i+T_{ij,j}+p_i\}\,\psi_i^{(n)}+\{D_{i,i}\}\,\varphi^{(n)}\,d\Omega=0, \tag{2.98}$$

where the volume equilibrium equations, including the loading, are projected on the corresponding components of the eigenfunctions. Following a similar procedure as that leading to Eq. (2.90) and described in Sect. 2.5.4, and since the mechanical and electrical loading at the boundary must now be taken into account, the previous expression becomes

$$-\int_\Omega \rho\ddot{w}_j\psi_j^{(n)}\,d\Omega-\int_\Omega T_{ij}S_{ij}^{(n)}-D_jE_j^{(n)}\,d\Omega+\int_\Omega p_j\psi_j^{(n)}\,d\Omega+\int_\Sigma \underline{\tau}_j\psi_j^{(n)}+$$
$$+\underline{\varrho}\varphi^{(n)}\,d\Sigma=0, \tag{2.99}$$

where $\underline{\varrho}=D_in_i$ is the surface charge on the boundary of the piezoelectric. Taking into account the modal decomposition (2.97), the orthogonality properties (2.95) and (2.96), the previous equation simplifies to

$$\ddot{\eta}_n+\omega_n^2\eta_n=\int_\Omega p_j\psi_j^{(n)}\,d\Omega+\int_\Sigma \underline{\tau}_j\psi_j^{(n)}+\underline{\varrho}\varphi^{(n)}\,d\Sigma, \tag{2.100}$$

which is equivalent to Eq. (2.56), given the new definition for the generalized modal forces, namely

$$\alpha_n=\int_\Omega p_j\psi_j^{(n)}\,d\Omega+\int_\Sigma \underline{\tau}_j\psi_j^{(n)}+\underline{\varrho}\varphi^{(n)}\,d\Sigma, \tag{2.101}$$

where we can notice the new term accounting for the electric energy supplied to the piezoelectric. On the nonelectroded part of the piezoelectric,

$\Sigma_{\bar{e}}$, there is obviously no contribution. On the electrodes, Σ_e, since the potential is constant, one has

$$\int_{\Sigma_e} \underline{\varrho} \varphi^{(n)} d\Sigma = \left(\int_{\Sigma_e} \underline{\varrho} d\Sigma \right) \varphi^{(n)} = -Q\varphi^{(n)}$$

where Q is the charge supplied by a generator; the minus sign results from the fact that, at the boundary separating the piezoelectric from the electrode, the charges are opposite. It follows that the electric charge is considered as a generalized force, and as such, open circuit electrical conditions (enforcing $Q = 0$) must be considered for the modal analysis. However, this implies that the generator is imposing a current, which is not often encountered in practice.

2.5.5.2 Imposed voltages

Alternatively, one could consider the electric displacement as the electrical generalized coordinate, so

$$\boldsymbol{W}(\boldsymbol{x}, t) = \begin{bmatrix} \boldsymbol{w}(\boldsymbol{x}, t) \\ \mathbf{D}(\boldsymbol{x}, t) \end{bmatrix} = \begin{bmatrix} \boldsymbol{\psi}^{(k)}(\boldsymbol{x}, t) \\ \mathbf{D}^{(k)}(\boldsymbol{x}) \end{bmatrix} \eta_k(t) = \boldsymbol{\Psi}_D^{(k)}(\boldsymbol{x})\eta_k(t). \tag{2.102}$$

Starting from the modified functional

$$\mathcal{P}'_{\psi_D^{(n)}}(\boldsymbol{W}) = \int_{\Omega} \{-\rho \ddot{w}_i + T_{ij,j} + p_i\} \psi_i^{(n)} - \{E_i + \varphi_{,i}\} D_i^{(n)} d\Omega = 0,$$

which now uses condition (2.69) involving the dual field $\boldsymbol{E}$, and being by construction null, the previous developments can be repeated, giving

$$\ddot{\eta}_n + \omega_n^2 \eta_n = \int_{\Omega} p_j \psi_j^{(n)} d\Omega + \int_{\Sigma} \underline{\tau}_j \psi_j^{(n)} d\Sigma - v \int_{\Sigma_e} \mathbf{D}^{(n)} \cdot \boldsymbol{n} d\Sigma. \tag{2.103}$$

Considering $\boldsymbol{D}$ as part of the generalized coordinates means that the generalized forces are the voltages, and so for the modal analysis the electrodes connected to a voltage generator should be considered short-circuited.

The following section now addresses more concrete aspects by narrowing the scope to thin ceramics which are usually encountered in practice.

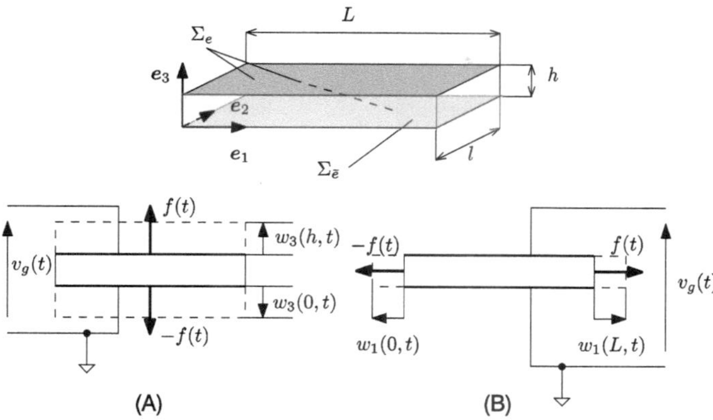

Figure 2.7 Thin ceramics in longitudinal (A) and transverse (B) coupling modes. They are both electroded on the lower and upper faces. The lower electrodes are grounded, and a voltage $v(t)$ is applied to the remaining electrodes. The mechanical load is supposed to be applied solely on the faces relevant to the coupling used.

2.6. Practical considerations

2.6.1 Modeling thin ceramics

In this part, we discuss the simplification of the general models that have been developed so far. The main assumption that will be used is that a dimension of the ceramic is small compared to the other. This is motivated by the trend of the available ceramics: to reduce the voltage applied, the manufacturers produce thin ceramics, electroded on the larger surfaces. Two cases widely encountered are represented in Fig. 2.7 where the longitudinal and transverse couplings are exploited. Since the potentials are imposed on the upper and lower faces, and moreover no free charges can appear on the nonelectroded faces, it is legitimate to consider that the electric field and the electric induction field only have a nonzero component along direction $\boldsymbol{e}_3$. The two cases are now examined in order to make the electric part of the modal force found in Eq. (2.103) more explicit.

2.6.1.1 Longitudinal vibrations

The mechanical problem is simplified by considering that only $T_3 \neq 0$ and $S_3 \neq 0$, it will be further assumed that they are only functions of x_3 and t. The equations for the problem are summed up below.

Equilibrium and material

Mechanic		Electric	
$T_{3,3} = -\rho \ddot{w}_3$	(2.104a)	$D_{3,3} = 0$	(2.105a)
$T_3 = c_{33}^D S_3 - h_{33} D_3$	(2.104b)	$E_3 = -h_{33} S_3 + \beta_{33}^S D_3$	(2.105b)
$S_3 = w_{3,3}$	(2.104c)	$E_3 = -\phi_{,3}$	(2.105c)

Boundary conditions

Mechanic		Electric	
$w_3(0, t) = 0$ or		$\phi(0, t) = 0$	(2.107a)
$\int_\Sigma T_3(0, t) d\Sigma = -f(t)$	(2.106a)	$-\int_\Sigma D_3(0, t) d\Sigma = Q(t)$	(2.107b)
$w_3(h, t) = 0$ or		$\phi(h, t) = v(t)$	(2.107c)
$\int_\Sigma T_3(h, t) d\Sigma = f(t)$	(2.106b)	$\int_\Sigma D_3(h, t) d\Sigma = -Q(t)$	(2.107d)

Taking into account the assumptions on the fields, Eq. (2.105a) implies that D_3 is only a function of time. By direct integration with respect to x_3 of Eq. (2.105b), one can deduce that the electric potential is

$$\phi(x_3, t) = h_{33} w_3(x_3, t) + \beta_{33}^S D_3(t)\, x_3 + c, \tag{2.108}$$

c being an arbitrary constant. Considering now the modal analysis, one puts $v(t) = 0$ meaning that the ceramic is short-circuited. Replacing $\psi_3^{(p)}(x_3)$ and $D_3^{(p)}$ into the previous result and solving Eqs. (2.107a) and (2.107c) in the case of the short-circuit yields

$$D_3^{(p)} = \frac{h_{33}}{\beta_{33}^S} \frac{\psi_3^{(p)}(h) - \psi_3^{(p)}(0)}{h}. \tag{2.109}$$

Denoting $\Delta\psi_3^{(p)} = \psi_3^{(p)}(h) - \psi_3^{(p)}(0)$, the contribution of the electric force to a mode is then simply

$$-v(t) \int_{\Sigma_e} D_3^{(p)} d\Sigma = \kappa_l^{(p)} v(t), \quad \kappa_l^{(p)} = -\frac{h_{33}}{\beta_{33}^S} \mathcal{S} \frac{\Delta\psi_3^{(p)}}{h}, \tag{2.110}$$

where $\mathcal{S} = L \times l$ is the surface of the electrode.

To highlight the reversibility of the piezoelectric effect, Eq. (2.108) is now considered when a voltage $v(t)$ is applied. The electric induction is then

$$D_3(t) = -\frac{1}{\beta_{33}^S} \frac{v(t)}{h} + \frac{h_{33}\mathcal{S}}{\beta_{33}^S h} \Delta w_3(t) \tag{2.111}$$

where $\Delta w_3 = w_3(h,t) - w_3(0,t)$. Since the charge supplied by the generator is $Q = -\int_{\Sigma_e} D_3 d\Sigma$, and noticing that $\Delta w_3(t) = \Delta\psi_3^{(p)}\eta_p(t)$, one can write

$$Q(t) = \frac{\mathcal{S}}{\beta_{33}^S}\frac{V}{h} - \frac{h_{33}\mathcal{S}}{\beta_{33}^S h}\left(\Delta\psi_3^{(p)}\eta_p(t)\right),$$

which gives the current

$$i(t) = C_0\dot{v}(t) + \kappa_l^{(p)}\dot{\eta}_p(t) \tag{2.112}$$

with the blocked capacitance $C_0 = \frac{\mathcal{S}}{\beta_{33}^S h}$. This last result shows that the total current is due to two contributions:

- the capacitive current – basically, the electroded dielectric is forming a capacitor C_0 which could be measured when applying a voltage while blocking the displacements;
- the motional current, $i_m(t) = \kappa_l^{(p)}\dot{\eta}_p(t)$, which results from the mechanical to electrical conversion and involves the modal velocities.

The careful reader might have noticed that the electromechanical conversion factor $\kappa_l^{(p)}$ is the same for the piezoelectrically induced modal force, and the motional current. This is, of course, a direct consequence of the reversibility of the piezoelectric coupling.

2.6.1.2 Transverse vibrations

The treatment is similar. Regarding the mechanical assumptions, T_1 and S_1 are now supposed to be the only nonzero components, and furthermore they are now supposed to depend on x_1 and t. So the problem is now stated by the equations:

Equilibrium and material

Mechanic		Electric	
$T_{1,1} = -\rho\ddot{w}_1$	(2.113a)	$D_{3,3} = 0$	(2.114a)
$T_1 = c_{11}^D S_1 - h_{31}D_3$	(2.113b)	$E_3 = -h_{31}S_1 + \beta_{33}^S D_3$	(2.114b)
$S_1 = w_{1,1}$	(2.113c)	$E_3 = -\phi_{,3}$	(2.114c)

Boundary conditions

Mechanic		Electric	
$w_1(0,t) = 0$ or		$\phi(0,t) = 0$	(2.116a)
$\int_\Sigma T_1(0,t)d\Sigma = -f(t)$	(2.115a)	$-\int_\Sigma D_3(0,t)d\Sigma = Q(t)$	(2.116b)
$w_1(L,t) = 0$ or		$\phi(l,t) = v(t)$	(2.116c)
$\int_\Sigma T_1(L,t)d\Sigma = f(t)$	(2.115b)	$\int_\Sigma D_3(l,t)d\Sigma = -Q(t)$	(2.116d)

Equation (2.114a) implies that D_3 depends only on x_1 and t. For the modal analysis, the electrodes are short-circuited, and one can verify that

$$D_3^{(p)}(x_1) = \frac{h_{31}}{\beta_{33}^S} S_1^{(p)}(x_1). \tag{2.117}$$

Then it is possible to calculate the modal force piezoelectrically produced, remembering that $S_1^{(p)} = \psi_{1,1}^{(p)}$:

$$-v(t) \int_{\Sigma_e} D_3^{(p)} d\Sigma = -v(t) \frac{h_{31} l}{\beta_{33}^S} \Delta \psi_1^{(p)}. \tag{2.118}$$

So once again, piezoelectric contribution to the modal force for a particular mode is simply proportional to the voltage, the proportionality coefficient being determined by the material properties, geometry and considered mode shape. Here, this coefficient is identified directly from Eq. (2.118) as

$$\kappa_t^{(p)} = -\frac{h_{31} l}{\beta_{33}^S} \Delta \psi_1^{(p)}. \tag{2.119}$$

Regarding the current, the reasoning is similar to that before. First, the charge on one of the electrodes is calculated. Considering Eq. (2.116d) with

$$D_3(x_1, t) = -\frac{v(t)}{\beta_{33}^S h} + \frac{h_{31}}{\beta_{33}^S} S_1(x_1, t),$$

which results from the direct integration of Eq. (2.114b) along direction $\boldsymbol{e}_3$, yields the charge

$$Q(t) = C_0 v(t) - \frac{h_{31} l}{\beta_{33}^S} \Delta w_1(t) \tag{2.120}$$

with $\Delta w_1(t) = w_1(L, t) - w_1(0, t)$. First, one can observe that the blocked capacitance has the same expression for the longitudinal and transversal cases. This is consistent with the previous comment: the origin of the equivalent capacitor is essentially the same in both cases if the displacements are blocked. Now, the current is expressed by taking the time derivative of the charge, and the modal expansion of the displacement, $w_1(x_1, t) = \psi_1^{(p)}(x_1)\eta_p(t)$, is also considered to reveal the modal velocities:

$$i(t) = C_0 \dot{v}(t) + \kappa_t^{(p)} \dot{\eta}_p(t).$$

So it can be concluded that from the electrical point of view there are no qualitative differences between the two couplings, and that the piezoelectric conversions are in both cases proportional to the voltage applied. The electromechanical conversion coefficient κ_t depends on the material, geometry and total elongation of the considered mode p.

Example 3 Transient response of piezoelectric ceramic subjected to a voltage

The thin ceramic with longitudinal coupling described by Eqs. (2.104a)–(2.107d) is supposed to be clamped on one of its faces (at $x_3 = 0$) the other being free, and mechanically loaded by a force on its free face at $x_3 = h$. Hence, the relevant mechanical boundary conditions are $w_3(0, t) = 0$ and $\int_\Sigma T_3(h, t) = 0$. The electrodes are on the horizontal faces, the lower one is supposed to be grounded. The ceramic is supposed to be at rest, initially undeformed, so $f(t) = 0$ $v(t) = 0$ for $t < 0$. At $t = 0$, a voltage is applied, so the potential of the upper electrode is supposed to be instantly $\phi(h, t) = v(t)$.

First, the mechanical equations are simplified considering Eqs. (2.105a), (2.104b) and (2.104a). Indeed, the dynamic equation reduces to

$$\rho \ddot{w}_3(x_3, t) = c_{33}^D w_{3,33}(x_3, t).$$

Earlier, the electric induction was related to the mode shape by Eq. (2.109) and in a general case by (2.111). This result is used to reformulate the mechanical boundary condition (2.106b). Considering that $f(t)$ is the result of a uniformly distributed force density $f(t) = \underline{\tau}_3 S$ on the surface, and replacing D_3, thanks to Eq. (2.111), the mechanical boundary conditions are now

$$c_{33}^D w_{3,3}(h, t) - \frac{h_{33}^2}{\beta_{33}^S} w_3(h, t) = \frac{f(t)}{S} - \frac{h_{33}}{\beta_{33}^S} \frac{v(t)}{h}. \tag{2.121}$$

In a similar way, the free vibrations' problem is modified. Using Eq. (2.109), the mode shapes are the solution of the modified set of equations:

Equilibrium		Boundary conditions	
$\psi_3^{(p)}(x_3) = \omega_p^2 \psi_{3,33}^{(p)}(x_3)$	(2.122)	$\psi_3^{(p)}(0) = 0$	(2.124)
$\omega_p^2 = \frac{\rho}{\underline{c}_{33}^D} \beta_p^2$	(2.123)	$\underline{c}_{33}^D \psi_{3,3}^{(p)}(h) = \frac{h_{33}^2}{\beta_{33}^S} \frac{\psi_3^{(p)}(h)}{h}$	(2.125)

where $\underline{c}_{33}^D = c_{33}^D - \frac{h_{33}^2}{\beta_{33}^S}$.

For these electrical boundary conditions, charges can flow from one electrode to the other, and a depolarizing electrical field can develop in the ceramic. This causes an induced stress that appears on the right-hand side of condition (2.125). The solution for the displacement is similar to that of Example 2, p. 62, except that, due to this electrically induced stress, the condition for the eigenvalues is now written as

$$\beta_p \cos(\beta_p h) - \frac{h_{33}^2}{c_{33}^D \beta_{33}^S} \sin(\beta_p h) = 0. \tag{2.126}$$

At this point, the eigenvalues must computed numerically. The normalization factor B_n is deduced from

$$B_p^2 \int_0^h \rho \mathcal{S} \left(\psi_3^{(p)}\right)^2 dx_3 = 1 \Leftrightarrow B_p^2 \rho \mathcal{S} \left(\frac{h}{2} - \frac{\sin(\beta_p h)}{4\beta_p}\right) = 1,$$

which means that the modal masses are not constant anymore.

Once the mode shapes and the resonant frequencies are known, the general problem can be casted in the modal space.

The modal forces are calculated thanks to the expressions found in (2.103), with $p_3 = 0, \underline{\tau}_3 = \frac{f(t)}{\mathcal{S}}$ and $D_3^{(p)}$ given by Eq. (2.109):

$$\alpha_p(t) = \int_\Sigma \frac{f(t)}{\mathcal{S}} \psi_3^{(p)}(h) d\Sigma - v(t) \int_\Sigma \frac{h_{33}}{\beta_{33}^S} \frac{\psi_3^{(p)}(h)}{h} d\Sigma$$

$$\Leftrightarrow \alpha_p(t) = F(t) B_p \sin(\beta_p h) + \kappa_l^{(p)} v(t), \tag{2.127}$$

with $\kappa_l^{(p)} = -\frac{h_{33}}{\beta_{33}^S} \frac{B_p \sin(\beta_p h)}{h}$ according to the result (2.110). The Laplace transform for the displacement is thus given by

$$W_3(x_3, s) = \sum_{k=1}^{\infty} B_k^2 \left(F(s) - \frac{h_{33}}{\beta_{33}^S} \mathcal{S} \frac{V(s)}{h}\right) \frac{\sin(\beta_k h)}{s^2 + 2\xi_k \omega_k s + \omega_k^2} \sin(\beta_k x_3). \tag{2.128}$$

To evaluate the influence of the piezoelectric effect, we consider a standard ceramic, for which

h_{33} ($\mathrm{V \cdot m^{-1}}$)	c_{33}^E ($\mathrm{N \cdot m^{-2}}$)	β_{33}^S ($\mathrm{m \cdot F^{-1}}$)	ρ ($\mathrm{kg \cdot m^{-3}}$)
23.5×10^8	12.1×10^{10}	1.37×10^8	7750

Assuming a thickness $h = 1$ mm, the resonant frequency for the first mode is nearly 80.9 kHz, while if we consider the solution $\beta_1 h = \frac{\pi}{2}$ found for the non-

piezoelectric rod, the resonant frequency is estimated to be 81 kHz, so the effect of the induced stress is a second-order effect.

2.6.2 Equivalent models

The main results of this theoretical discussion are:

- Any piezoelectric device can be modeled as an infinite set of damped oscillators with one degree of freedom.
- Each oscillator displacement represents the response of a given mode shape, or modal coordinate.
- Each oscillator is excited by a linear combination of the force and voltage applied to the device.
- The actual displacement (space and time function) is obtained by superposing the individual responses of the oscillators, that is, by summing the modal shapes (space functions) multiplied by their respective modal coordinates (time functions).
- The piezoelectric effect is reversible, so the current is the superposition of a capacitive current (the piezoelectric ceramic is basically two electrodes separated by a dielectric) and a motional current which is proportional to the velocity at the boundary of the ceramic.
- Because the piezoelectric effect is reversible, the electrical-to-mechanical and mechanical-to-electrical conversion coefficients are necessarily equal.

The development presented in this example assumes that the ceramic is thin in the polarization direction, and as a result, the electric field and the electric displacement field only have nonzero component in the $\boldsymbol{e}_3$ direction. In practice, ceramics are indeed thin (in order to reach high values of the electric field while reducing the voltage applied), and this simplification is well justified.

The examples that were studied are helpful to predict the main dynamic parameters of the device according to some geometry and material properties. In many cases, though, the engineer is not interested in the design, but needs a way to rapidly build a dynamic model of a given device, for simulation purposes, for instance. The observations above suggest using the modal representation because the problem is reduced to simple oscillators, and the piezoelectric conversion can be seen as a kind of electromechanical ideal "transformer".

However, in the equations that were considered so far, the mode shapes were normalized with respect to the vibrating mass of the mode. This normalization was introduced because the mode shapes – which are from the mathematical point of view eigenvectors – are defined up to a multiplicative constant. Hence, the normalization considered so far was "designed" to simplify the vibrating mass of a mode in order to simplify the mathematical development. By doing so, the units for the normalization factors are $kg^{-\frac{1}{2}}$, while the modal coordinates are in $kg^{\frac{1}{2}} \cdot m$, in other words, they have no intrinsic physical meaning, although their product is a displacement.

Rather than using such normalization, it is often more interesting to normalize so that the maximum of the mode is one (with no units), for instance, in the case of experimental measurements of the mode shapes. Let $\boldsymbol{\psi}$ be a nonnormalized mode, then such a normalization is given by

$$\boldsymbol{\Psi}(\boldsymbol{x}) = \frac{\boldsymbol{\psi}(\boldsymbol{x})}{\hat{\psi}} \tag{2.129}$$

where $\hat{\psi} = \max(\|\boldsymbol{\psi}(\boldsymbol{x})\|)$ for $\boldsymbol{x} \in \Omega$. The normalized modal shapes are now unitless, and as a result, the modal coordinates are now physical displacements and are actually the maximum amplitude of a given mode subjected to a given modal force. To relate this normalization to the previous theory, we should now write the orthogonality relation as

$$\int_{\Omega} \rho \Psi_i^{(n)} \Psi_i^{(p)} d\Omega = \delta_{np} m_p \tag{2.130}$$

in kilograms, which is the vibrating mass of the mode. Then, the following equality holds for the potential energy:

$$\int_{\Omega} T_{ij}^{(n)} S_{ij}^{(p)} - D_i^{(n)} E_i^{(p)} d\Omega = \delta_{np} m_p \omega_p^2 = \delta_{np} k_p, \tag{2.131}$$

yielding the mode rigidity.

This normalization only introduces a scaling factor and so the projection method remains applicable, except that now the modal equations are

$$m_p \ddot{w}_p + k_p w_p = f_r^{(p)} + f_p^{(p)}. \tag{2.132}$$

Here $w_p(t)$ denotes the actual amplitude of a mode, and $f_r^{(p)}$ is the projection of the "mechanical" forces

$$f_r^{(p)} = \int_{\Omega} p_k \Psi_k^{(p)} d\Omega + \int_{\Sigma} \tau_k \Psi_k^{(p)} d\Sigma, \tag{2.133}$$

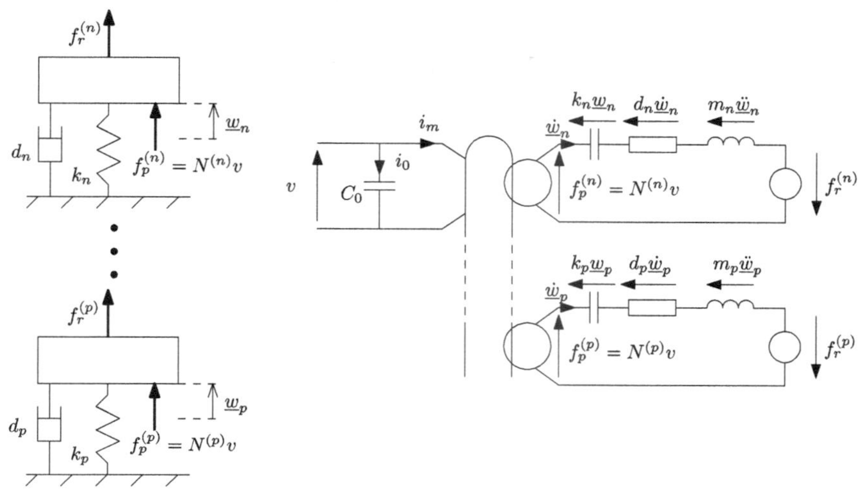

Figure 2.8 Equivalent representation of a piezoelectric device: the dynamics can be represented as an infinite set of elementary oscillators with unit mass for a convenient normalization (left); an equivalent electric circuit can also be considered where the mechanical modes are transposed to electrical modes of an infinite set of second-order electric circuits (right).

and $f_p^{(p)}$ is the projection of the "piezoelectric" force, calculated, for instance, by

$$f_p^{(p)} = -v \int_{\Sigma_e} \mathbf{D}^{(p)} \cdot \boldsymbol{n} d\Sigma = N^{(p)} v, \tag{2.134}$$

where $N^{(p)}$ is named the *force factor* or the *coupling factor.*

On the electrical side, only the motional current is affected by the change in the normalization. From the observation made previously, since to respect the reversibility the coupling factor has to be the same, the motional current will be written as

$$i(t) = C_0 \dot{v}(t) + N^{(p)} \dot{w}_p(t). \tag{2.135}$$

Several analogies can be deduced from Eqs. (2.132)–(2.135). The first one can adopt a mechanical point of view, where each mode is replaced by a single degree of freedom linear oscillator, with mass m_p, rigidity k_p and, to account for the losses, a damping d_p. They are excited by the modal projection f_p of the forces (2.134) and the piezoelectric forces (2.134) applied to the actual transducers. Such equivalent systems are represented in Fig. 2.8.

As the piezoelectric device is an electromechanical transducer, it can be similarly described by adopting an equivalent electric circuit. For this purpose, one assimilates the speeds to currents, and forces to voltages, and substitutes to each mechanical oscillator an equivalent series electrical oscillator featuring an inductance with value m_p, a capacitance with value $\frac{1}{k_p}$, and a resistor with value d_p. These electrical oscillators are the motional branches of the equivalent circuit. The schematic is completed by an "electromechanical transformer" that dispatches voltage v among the various motional branches. The transformer ratio relating the applied voltage to the modal piezeoelectric force is then $N^{(p)}$. Finally, the modal force is represented as a voltage generator.

The other side of the electromechanical transformer features the blocked capacitance C_0 in parallel, so the total current is the sum of the capacitive current and the motional currents as expressed by Eq. (2.135).

CHAPTER THREE

Modeling in a rotating reference frame

Contents

3.1. Causal modeling of a piezoelectric transducer 96
- **3.1.1** What is causal modeling? 96
- **3.1.2** Lumped parameter model of a piezoelectric transducer 97
- **3.1.3** EMR of a piezoelectric transducer 98
- **3.1.4** Simulation results 100

3.2. Introduction to the rotating reference frame 103

3.3. Modeling in the rotating reference frame 106
- **3.3.1** Modeling in steady state 106
  - *3.3.1.1 No-load operations 107*
  - *3.3.1.2 Operating point when loaded 108*
- **3.3.2** Dynamic modeling in the rotating reference frame 110
- **3.3.3** Simulation results 113

3.4. Modeling a traveling wave in a linear beam 114
- **3.4.1** The multimodal approach 115
- **3.4.2** Model of a beam with two exciters 116
- **3.4.3** Simulation results 120
- **3.4.4** Modeling of a Traveling Wave Ultrasonic Motor in the rotating reference frame 123
  - *3.4.4.1 The model with two independent phases 124*
  - *3.4.4.2 EMR of the TWUM with two independent phases 130*
  - *3.4.4.3 Simplified model 133*

Abstract

The equations which describe the evolution of the state variables in a piezoelectric actuator are organized in order to obtain a causal modeling. We then describe this model into a rotating reference frame. Simulations are presented, that help to understand the behavior of ultrasonic actuators: the Langevin transducer and the Traveling Wave Ultrasonic Motor.

Keywords

causal modeling, rotating reference frame, finite element model, equivalent mass-spring system, energetic macroscopic representation, traveling wave

Piezoelectric Actuators: Vector Control Method
https://doi.org/10.1016/B978-0-12-814186-1.00007-7

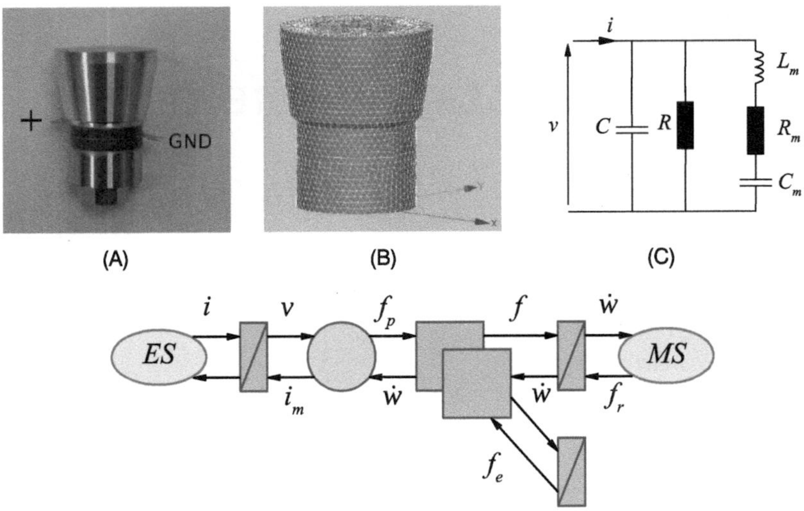

Figure 3.1 Three ways to model a Langevin transducer: (A) the device, (B) finite element model, (C) electrical equivalent circuit, and (D) causal representation.

3.1. Causal modeling of a piezoelectric transducer

3.1.1 What is causal modeling?

Engineers and researchers have several types of models at their disposal to describe the behavior of the system they study (See Fig. 3.1.). The level of accuracy of their model depends on the purpose they are used for. For example, a motor designer will use a finite element model of the system, which precisely calculates the motor's output for a given set of inputs. This type of model is very accurate, it can take into account complex geometries and the properties of the material used, but the accuracy is obtained at the expense of complexity: it contains millions of variables, and the resolution is only possible with a computer. Lumped parameter models, on the other hand, are much less complicated models. They are based on a set of assumptions and parameters derived from a combination of the material properties and geometry of the motor. Because they use a limited number of state variables to describe the system's behavior, a formal resolution can sometimes be calculated, and it is always faster than a finite element model. This type of model is often used to study the interaction of the system with its environment, for instance, in the case of a motor and its power electronic.

When an engineer wants to design a controller for his motor, what he needs is a model for the control. For that purpose, he will lay down

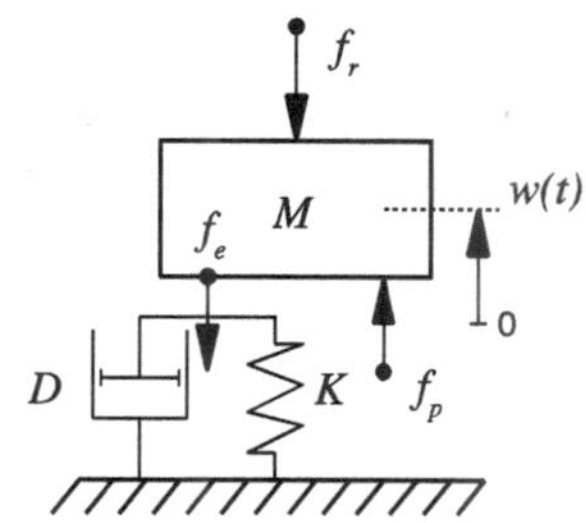

Figure 3.2 Equivalent mass–spring system of a Langevin transducer supplied at a pulsation close to the resonance of a chosen mode. By convention, the force of the environment f_r is reversed compared to Chap. 2.

new assumptions which help reduce the model's size and complexity; this is achieved by removing phenomena whose dynamics are faster than the bandwidth of the closed-loop control, as explained in Chap. 1. Hence, the accuracy of a model for the control is not very important. Of utmost importance is to describe the *causality* in the system, i.e., the relations between the causes and effects. This way, a control can be proposed by inverting the model, and the controllers will compensate for the discrepancies between the model and real system.

The purpose of this chapter is to present the construction of a causal model of a piezoelectric transducer, step by step, and to simulate it using a simulation software. ®MATLAB Simulink is used in this book as it is a very popular software suite, and we will present the methodology to obtain a simulation worksheet in Simulink from the model.

3.1.2 Lumped parameter model of a piezoelectric transducer

Most of the piezoelectric transducers exploit a single vibration mode to convert electricity into mechanical power. This has been shown for the benders, Langevin transducers and many other devices in Chap. 1. Therefore, in this section, we will consider a piezoelectric resonator which exploits a particular vibration mode (p). Moreover, we will assume that the vibration pulsation ω is very close to the resonant pulsation $\omega^{(p)}$. As a result, the contributions of all the other modes are negligible, and therefore, these unused modes can be removed from the model depicted in Fig. 2.8, leading to a simple mass–spring system for the mechanical part, as depicted in Fig. 3.2. For the sake of clarity, the equivalent mass m_p, stiffness k_p, damping d_p, and the force factor $N^{(p)}$ calculated from Eqs. (2.130)–(2.134) from Chap. 2 are now denoted M, K, D, and N, respectively.

This mass–spring system leads to Eq. (3.1), which describes the evolution of the displacement w by a second-order differential equation, forced by the modal forces f_p and f_r[1]:

$$M\ddot{w} + D\dot{w} + Kw = f_p - f_r. \tag{3.1}$$

In addition to Eq. (3.1), the energy conversion process, which occurs due to the piezoelectric material, leads to

$$f_p = Nv, \quad i_m = N\dot{w}, \tag{3.2}$$

where v is the supplied voltage and i_m is the *motional current*. The capacitive behavior of the piezoelectric material is then described by

$$i = C_0 \frac{\mathrm{d}v}{\mathrm{d}t} + i_m. \tag{3.3}$$

Eqs. (3.1)–(3.3) allow calculating the evolution of the displacement $w(t)$. The next section deals with the representation of this modeling using the energetic macroscopic representation (EMR).

3.1.3 EMR of a piezoelectric transducer

The causality principle used in EMR leads us to remove the derivatives in the model, and use integral forms instead. For this purpose, Eq. (3.1) gives

$$\ddot{w} = \frac{1}{M}\left(f_p - f_r - D\dot{w} - Kw\right), \tag{3.4}$$

leading to

$$\dot{w} = \frac{1}{M}\int \left(f_p - f_r - D\dot{w} - Kw\right)\mathrm{d}t. \tag{3.5}$$

It is convenient to introduce $f_e = (D\dot{w} + Kw)$ and $f = f_p - f_e$ in Eq. (3.5) in order to obtain its causal form

$$\dot{w} = \frac{1}{M}\int \left(f - f_r\right)\mathrm{d}t. \tag{3.6}$$

[1] In Chap. 2, the sign convention of f_r is positive, meaning that if f_r is positive, w expands; in this chapter, the sign convention of f_r is reversed, which means that if f_r is positive, w contracts. Therefore, by convention, if $f_r\dot{w} > 0$, the environment receives energy from the transducer.

The energetic macroscopic representation (EMR)

The energetic macroscopic representation decomposes a system into basic subsystems or blocks, and connects them according to the action and reaction principle by exchange variables represented by arrows (Bouscayrol et al., 2013). The product of action and reaction variables gives rise to the power flowing in the system. EMR uses the principle of causality to determine the direction of arrows: when a variable is an integral function of others, it becomes the output of a block. Moreover, by principle, it is forbidden to connect two outputs or two inputs together. The basic blocks are described in the following table.

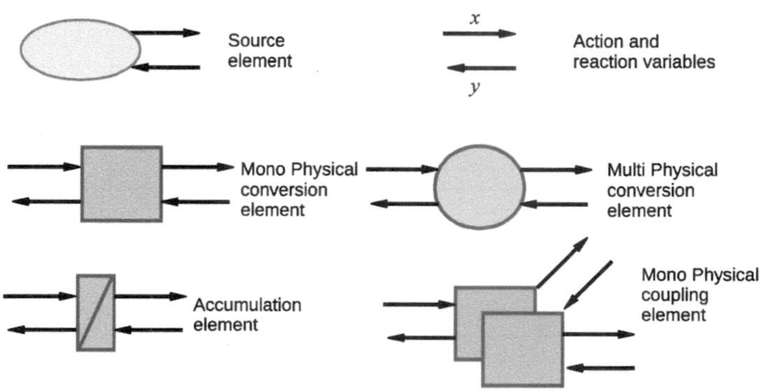

From Eq. (3.5), we can deduce that $\dot{w}$ is an output, because it is an integral function of f and f_r, which are inputs. Therefore, a convenient representation of this equation is depicted in Table 3.1, in which we have also summarized the other equations and their representations.

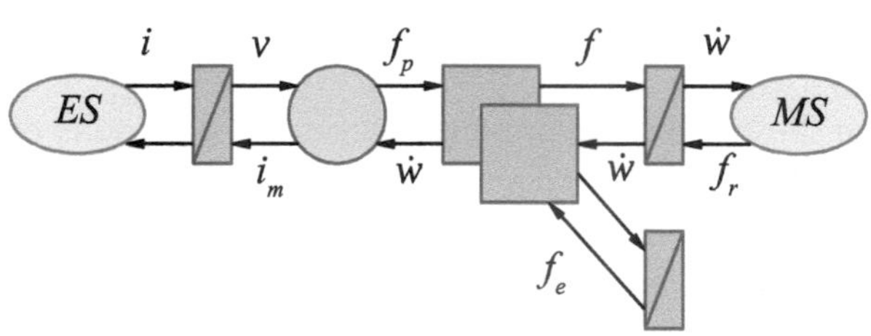

Figure 3.3 An energetic macroscopic representation of a Langevin transducer.

Assembling all the blocks of the EMR gives the representation of the model of the Langevin transducer in Fig. 3.3; we add two source blocks (represented by ellipses), which represent the mechanical load to the actuator (*i.e.*, the relation between f_r and $\dot{w}$) and the power supply (current i).

Table 3.1 Causal modeling and its representation of a Langevin transducer.

Equation	Representation	Simulink blocks
$\dot{w} = \frac{1}{M}\int (f - f_r)\,\mathrm{d}t$	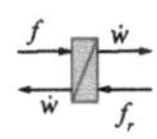	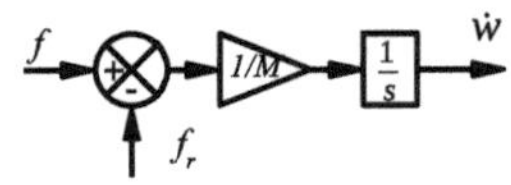
$f_e = D\dot{w} + K\int \dot{w}\mathrm{d}t$	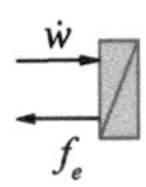	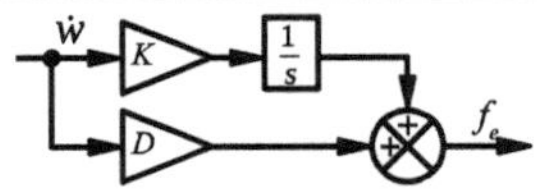
$v = \frac{1}{C_0}\int (i - i_m)\,\mathrm{d}t$	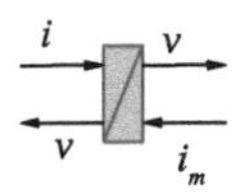	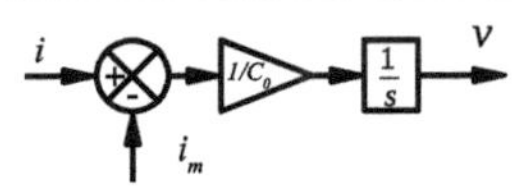
$f_p = Nv$ and $i_m = N\dot{w}$	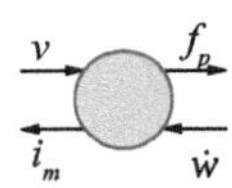	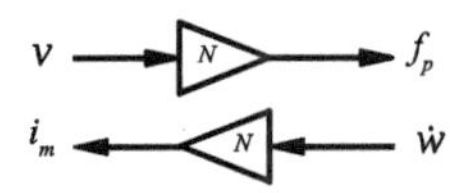
$f = f_p - f_e$	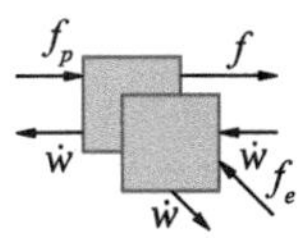	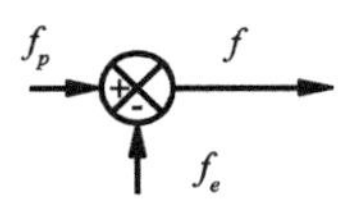

This process helps organize the equations, which can be difficult for large systems. Moreover, since each block of the EMR can be converted into Simulink blocks, as shown in Table 3.1, we have the simulation worksheet quite easily. And if we encapsulate these blocks into subsystems as those represented by the blocks of the EMR, we obtain an organized and clear Simulink worksheet. This methodology will be applied all over this book.

3.1.4 Simulation results

For the purpose of illustration, we present two simulation results of a typical Langevin transducer. The parameters used were: $M = 0.249$ kg, $D = 58.4$ Ns/m, $K = 7.64 \cdot 10^9$ N/m, $N = 0.45$ N/V, and $C_0 = 2.36$ pF. These parameters lead to a resonance pulsation of $\omega_0 = \sqrt{K/M} = 175\,160$ rad/s, which is equivalent to a resonance frequency of 27.9 kHz. The system is supposed to operate in vacuum, and no load is taken into account, $f_r = 0$.

For the first simulation try, voltage is applied to the system, with a constant peak amplitude of 100 V, but with a varying pulsation ω. At $t = 0$, the pulsation follows a step variation from $1.005 \times \omega_0$ to $1.001 \times \omega_0$, meaning that the system gets closer to its resonance: the vibration amplitude

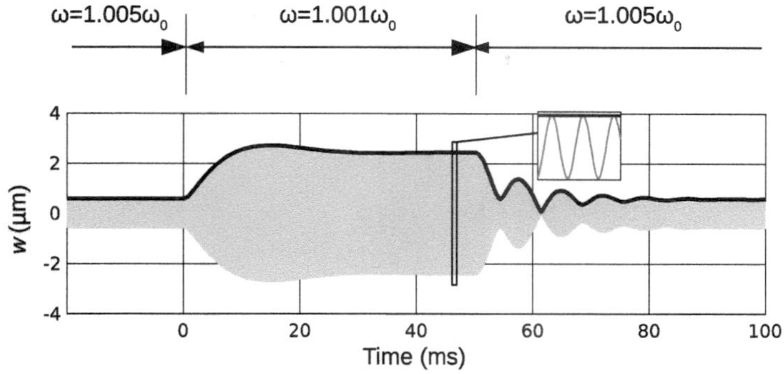

Figure 3.4 Response of the displacement to step variation of the pulsation; $w(t)$ is in grey, its peak amplitude is in black.

increases. At time $t = 50$ ms, an inverse step variation is imposed to the system, and the vibration amplitude decreases. The response is given in Fig. 3.4.

Hence, from $t = 0$ the vibration amplitude reaches its steady state in approximately 20 ms with a small overshoot. At $t = 50$ ms, the falling down of the vibration amplitude is longer, with oscillations. Evidently, the transitory behavior of the vibration amplitude is not the same at the falling edge as at the rising edge. This reveals a nonlinear behavior of the system besides variation of the pulsation. It should be emphasized here that the system is *not* nonlinear: Eqs. (3.1)–(3.6) *are* linear. It is the response of the system to variation of the pulsation of the voltage which is nonlinear. Therefore, the transfer function between the vibration amplitude and pulsation has poles which are not constant, and which depend on the operating point itself.

In the second try, the pulsation is kept constant, and the voltage amplitude is increasing from 20 to 100 V. We compare in this figure the transitory response at several pulsations: $\omega \in \{1.005 \times \omega_0, 1.0025 \times \omega_0, 1.001 \times \omega_0, \omega_0\}$. For the sake of clarity, only the peak amplitude of the vibration is represented. (See Fig. 3.5.)

Under these conditions also the transitory response of the system depends on the pulsation. The vibration amplitude oscillates before stabilizing, and the overshoot depends on the operating point. The case at resonance differs from the others, since there is no overshoot: actually, the response of the vibration amplitude is of the first-order type with a time constant equal to $2M/D$.

Transitory response at resonance

The transitory response of a resonating system, is particular when it is excited by a force at resonance. Let us consider a second-order type equation

$$M\ddot{w} + D\dot{w} + Kw = Nv(t). \tag{3.7}$$

If this system is excited at resonance, we write

$$v = V\cos(\omega_0 t), \tag{3.8}$$

with $\omega_0 = \sqrt{\frac{K}{M}}$. In order to calculate the evolution of w as a function of time t, we use the Laplace transforms of $v(t)$ and $w(t)$, namely $\widehat{V}(s)$ and $\widehat{W}(s)$, with s being the Laplace variable. Using the usual Laplace transform table, Eqs. (3.7) and (3.8) lead to

$$\widehat{V}(s) = \frac{s}{s^2 + \omega_0^2} V \tag{3.9}$$

and

$$\begin{aligned}\widehat{W}(s) &= \frac{s}{\left(s^2 + \omega_0^2\right)\left(s^2 + \frac{D}{M}s + \omega_0^2\right)} \frac{NV}{M} \\ &= \left[\frac{1}{s^2 + \omega_0^2} - \frac{1}{s^2 + \frac{D}{M}s + \omega_0^2}\right] \frac{NV}{D}.\end{aligned} \tag{3.10}$$

By using the inverse Laplace transforms, we can obtain the original function of $\widehat{W}(s)$, which is $w(t)$:

$$\begin{aligned}w(t) &= \left[\frac{1}{\omega_0}\sin(\omega_0 t) - \frac{1}{\omega_0} e^{-\frac{2M}{D}t}\sin(\omega_0 t)\right] \frac{NV}{D} \\ &= \frac{NV}{D\omega_0}\left(1 - e^{-\frac{2M}{D}t}\right)\sin(\omega_0 t).\end{aligned} \tag{3.11}$$

The evolution of the displacement is sinusoidal, while the vibration amplitude follows the function $\left(1 - e^{-\frac{2M}{D}t}\right)$. It is interesting to notice that this function represents the response of a first-order system to a step input. It can be shown that this behavior is obtained only if the working frequency is at resonance, while at other frequency, a second-order behavior is obtained (with overshoot, for instance): switching on and off the supply voltage periodically, and adjusting the frequency until the response of the vibration amplitude is of first-order is a good way to identify the value of the resonance frequency.

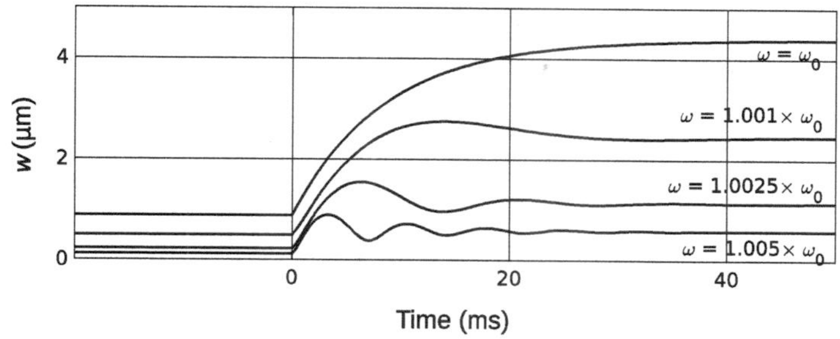

Figure 3.5 Response of the vibration amplitude to step variation of the voltage amplitude, for several pulsations.

The model's output is the displacement $w(t)$. This is why we had to calculate the peak amplitude (or the vibration amplitude) in Fig. 3.4 with an offline algorithm. This can be achieved with a peak detection method, for instance. The purpose of the modeling in the rotating reference frame is to easily calculate the dynamic evolution of the peak (vibration) amplitude directly. We will see that this new modeling has interesting properties for the control.

3.2. Introduction to the rotating reference frame

The variables introduced in Chap. 2 have a physical meaning: we find the voltage $v(t)$, current $i(t)$, and displacement $w(t)$. These variables can be measured with instruments: voltage and current sensors, laser interferometers, oscilloscopes, etc. Hence, the modeling presented in the same chapter gives the evolution of the variables as a function of time t. Since it is based on a reasonable number of assumptions, it is useful for design purposes. Moreover, it is suitable for studying nonlinear phenomena, like the introduction of second order harmonics in nonlinear transducers, because it can predict the evolution of the variables besides nonlinear parameter evolution.

In practice, however, the instantaneous evolution of the aforementioned variables does not really matter. In fact, most of the ultrasonic transducers are designed in such a way that harmonics are neglected, and the variables are considered sinusoidal functions of time. In these conditions, the operating point of the transducers is characterized by the amplitude of the current, voltage and vibration. This is why a modeling which captures the dynamics of the amplitude rather than the instantaneous values is necessary.

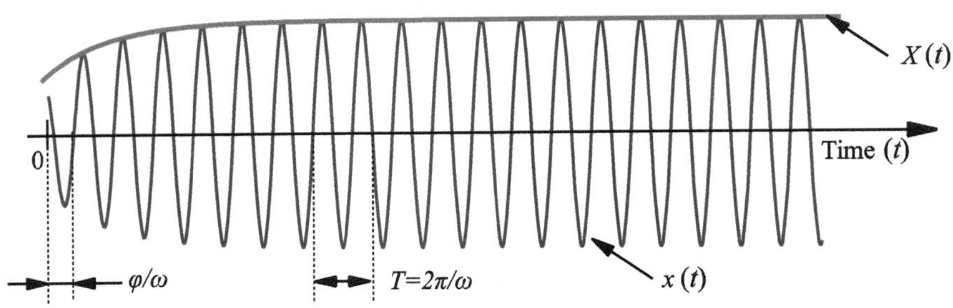

Figure 3.6 Representation of $x(t)$ as a function of time.

To develop such a model, we first assume that the variables are sinusoidal functions of time, and we write

$$x(t) = X(t)\sin(\omega t - \varphi) \tag{3.12}$$

where $x(t)$ is the variable, $X(t)$ is its amplitude, ω is the pulsation, and φ is the phase of x. A representation of $x(t)$ is given in Fig. 3.6: to each time t corresponds a point, the position of which on the y axis is $x(t)$. We also depicted $X(t)$ as the envelope of $x(t)$, $T = 2\pi/\omega$ the period of $x(t)$ and the phase φ.

A *complex phasor* is attached to $x(t)$; it is denoted by $\underline{x}(t)$, and it is a complex number. By definition, we write

$$x(t) = \Im\left(\underline{x}(t)\right) \tag{3.13}$$

where $\Im(\bullet)$ represents the imaginary part.

It is then convenient to define the complex amplitude $\underline{X} = Xe^{-j\varphi}$, such that

$$\underline{x} = \underline{X}e^{j\omega t}. \tag{3.14}$$

Substituting (3.14) into (3.13) gives

$$\begin{aligned}\Im\left(\underline{X}e^{j\omega t}\right) &= \Im\left(Xe^{-j\varphi}e^{j\omega t}\right)\\ &= \Im\left(Xe^{j(\omega t-\varphi)}\right)\\ &= X\sin(\omega t - \varphi) = x(t).\end{aligned} \tag{3.15}$$

Complex numbers are also vectors in a 2D plane. Hence, $\underline{x}$ is a vector in such a plane, the projection of which on the imaginary axis gives $x(t)$

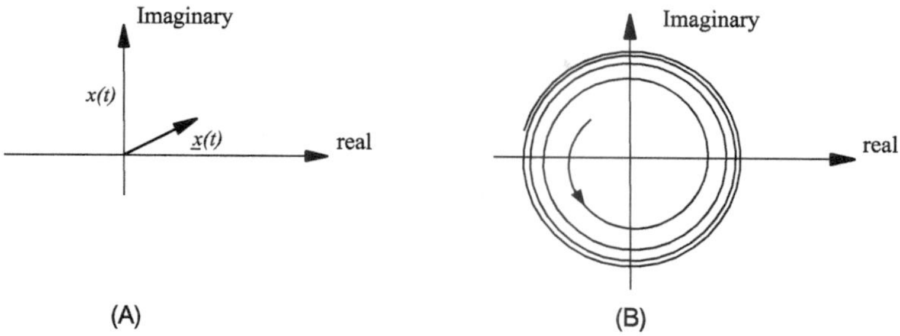

Figure 3.7 Representation of $\underline{x}(t)$ in the complex plane: (A) at time t, (B) evolution of the affix of the vector $\underline{x}$ corresponding to Fig. 3.6.

due to Eq. (3.13). Therefore, to each time t corresponds a vector $\underline{x}$, with length $X(t)$ and position equal to $(\omega t - \varphi)$ as depicted in Fig. 3.7 (A).

When $x(t)$ changes as depicted in Fig. 3.6, the position and length of vector $\underline{x}$ are also modified. Fig. 3.7 (B) shows the evolution of the extremity of $\underline{x}$ under this condition. It moves along a spiral. When X is constant, the length of $\underline{x}$ stays constant, and its extremity rotates because its phase $(\omega t - \varphi)$ still varies with time. Therefore, $\underline{x}$ is a *rotating vector*, and it is convenient to represent it in a *rotating reference frame*. For that purpose, we define a new reference frame, denoted by (d, q), which is orthonormal and has axis d aligned with $e^{j\omega t}$ as depicted in Fig. 3.8 (A); by definition of (d, q), vectors in this reference frame can be transformed to the original one by a rotation with an angle ωt.

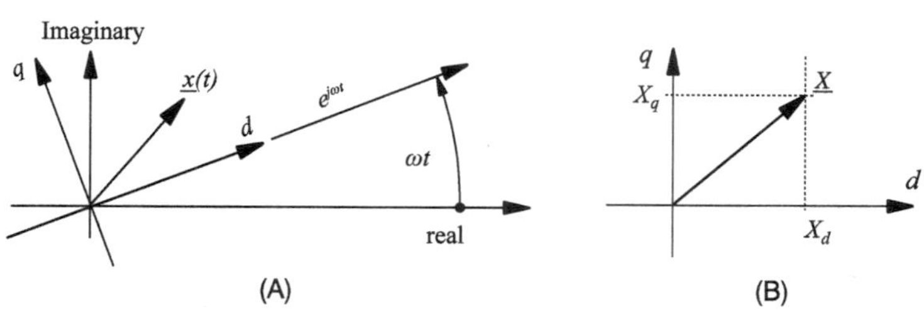

Figure 3.8 Representation of $\underline{x}(t)$ in a rotating reference frame (d, q): (A) definition of the reference frame (d, q), (B) coordinates of $\underline{X}$ in (d, q).

It is then obvious from Eq. (3.14) that $\underline{x}$ in the fixed reference frame and $\underline{X}$ in the (d, q) frame represent the same vector. It is then convenient to define X_d and X_q as the projections of $\underline{X}$ on axes d and q, respectively. By extension, X_d and X_q are also called the coordinates of $\underline{x}$ in the rotating reference frame.

It should be emphasized here that in steady state, when X is constant, the signal $x(t)$ is continuously evolving as a sinusoidal function of time, $\underline{x}$ is rotating in the fixed reference frame, while $\underline{X}$ is constant in the (d, q) frame. This property is important, and will be exploited in Chap. 4 which deals with the control of the Langevin transducer.

Usually, in a system, there is only one rotating reference frame. In most of the cases, we choose to align the reference frame with a specific vector. For instance, axis d can be aligned with voltage $\underline{v}$, meaning $V_q = 0$.

3.3. Modeling in the rotating reference frame

The modeling in the rotating reference frame consists in giving the evolution of the coordinates X_d and X_q of each variable. For that purpose, we admit that we can replace variable x by its complex phasor $\underline{x}$ in Eqs. (3.1); the general equations of the transducer are then given by

$$\begin{aligned} M\underline{\ddot{w}} + D\underline{\dot{w}} + K\underline{w} &= \underline{f}_p - \underline{f}_r, \\ \underline{i}_m &= N\underline{\dot{w}}, \\ \underline{v} &= \frac{1}{C}\int (\underline{i} - \underline{i}_m)\,\mathrm{d}t. \end{aligned} \tag{3.16}$$

In this section, we will first study the steady state of an ultrasonic transducer. Then, we will give its dynamic modeling.

3.3.1 Modeling in steady state

During steady state, the pulsation ω of the voltage, current and displacement is constant. Moreover, the amplitudes of these variables are also constant, and therefore, X_d and X_q of the variable $\underline{x}$ are constant. This is used to calculate the first derivative of $\underline{w}$:

$$\begin{aligned} \underline{\dot{w}} = \frac{\mathrm{d}\underline{w}}{\mathrm{d}t} &= \frac{\mathrm{d}\left(W_d + jW_d\right)e^{j\omega t}}{\mathrm{d}t} \\ &= \left(W_d + jW_d\right)\frac{\mathrm{d}e^{j\omega t}}{\mathrm{d}t} \\ &= \left(W_d + jW_d\right)j\omega e^{j\omega t}. \end{aligned} \tag{3.17}$$

For the sake of simplification, we introduce in this chapter the velocity $u(t) = \dot{w}$, and its complex phasor $\underline{u} = \left(U_d + jU_q\right)e^{j\omega t}$. The vibration velocity amplitude U is defined as $U = \| U_d^2 + U_q^2 \|$.

With these new variables, Eq. (3.17) leads to $U_q = \omega W_d$ and $U_d = -\omega W_q$. We then obtain for the steady state

$$\ddot{\underline{w}} = j\omega \left(U_d + jU_q\right) e^{j\omega t} \tag{3.18}$$

and

$$\underline{w} = \frac{U_d + jU_q}{j\omega} e^{j\omega t}. \tag{3.19}$$

By taking into account (3.18) and (3.19) in (3.16) leads to

$$\begin{aligned} &\left(\frac{K - M\omega^2}{j\omega}\right)\left(U_d + jU_q\right) e^{j\omega t} + D\left(U_d + jU_q\right) e^{j\omega t} \\ &= \left(F_{pd} + jF_{pq}\right) e^{j\omega t} - \left(F_{rd} + jF_{rq}\right) e^{j\omega t}. \end{aligned} \tag{3.20}$$

Since $e^{j\omega t} \neq 0$, Eq. (3.20) simplifies, and we write:

$$\left(\frac{K - M\omega^2}{j\omega}\right)\left(U_d + jU_q\right) + D\left(U_d + jU_q\right) = \left(F_{pd} + jF_{pq}\right) - \left(F_{rd} + jF_{rq}\right). \tag{3.21}$$

By splitting the real and imaginary part in Eq. (3.21) leads to:

$$\begin{aligned} \frac{K - M\omega^2}{\omega} U_q + DU_d &= F_{pd} - F_{rd}, \\ -\frac{K - M\omega^2}{\omega} U_d + DU_q &= F_{pq} - F_{rq}. \end{aligned} \tag{3.22}$$

The two equations in (3.22) represent the model of the transducer in the rotating reference frame, and for steady-state; we remark that the two axes are very coupled, and that ω acts as a nonconstant parameter. The next section details the operation of the transducer at no-load.

3.3.1.1 No-load operations

The study of a Langevin transducer often starts without any mechanical load attached to it, in order to identify some parameters, or observe its no-load behavior. Hence, in this part, the forces F_{rd} and F_{rq} are set to 0. Moreover, the voltage amplitude V is kept constant, and ω is varying; aligning the reference frame with the voltage leads to $V_q = 0$. Under these conditions, equations in (3.22) become:

$$\begin{aligned} &\frac{K - M\omega^2}{\omega} U_q + DU_d = NV, \\ -&\frac{K - M\omega^2}{\omega} U_d + DU_q = 0. \end{aligned} \tag{3.23}$$

We now look for the solutions (U_d, U_q) which satisfy (3.23). We first remark that if $U_q = 0$, then $U_d = 0$, too, and if $U_q \neq 0$, then we have

$$\frac{K - M\omega^2}{\omega} = D\frac{U_q}{U_d}. \tag{3.24}$$

Substituting (3.24) into (3.23) and simplifying by D leads to

$$\frac{U_q^2}{U_d} + U_d = \frac{NV}{D} \rightarrow U_q^2 + U_d^2 = \frac{NV}{D} U_d, \tag{3.25}$$

$$U_q^2 + \left(U_d - \frac{NV}{2D}\right)^2 = \left(\frac{NV}{2D}\right)^2. \tag{3.26}$$

Eq. (3.26) is the equation of a circle. Therefore, when sweeping ω of both sides of the resonance, the affix of the vector $\underline{U}$ follows a circle with

- center located at $\left(\frac{NV}{2D}, 0\right)$,
- radius equal to $\frac{NV}{2D}$,

as depicted in Fig. 3.9. Several properties for this circle can be deduced:

- $U_d > 0$ for each ω;
- if $\omega < \omega_0$ (resp. $\omega > \omega_0$) then Eq. (3.23) yields $U_q > 0$ ($U_q < 0$ resp.);
- at resonance, Eq. (3.23) yields $U_q = 0$;
- when ω increases, the operating point follows the circle in a clockwise direction.

Hence, resonance can be found as the pulsation at which length U is maximum, or, and this is equivalent for this system, at which $U_q = 0$.

3.3.1.2 Operating point when loaded

It is not possible to give a general form of f_r: it depends on the type of load. But we can give the load the form of a mass–spring system, since this corresponds to a large number of cases. Therefore, we will write

$$f_r(t) = \tilde{M}\ddot{w} + \tilde{D}\dot{w} + \tilde{K}w. \tag{3.27}$$

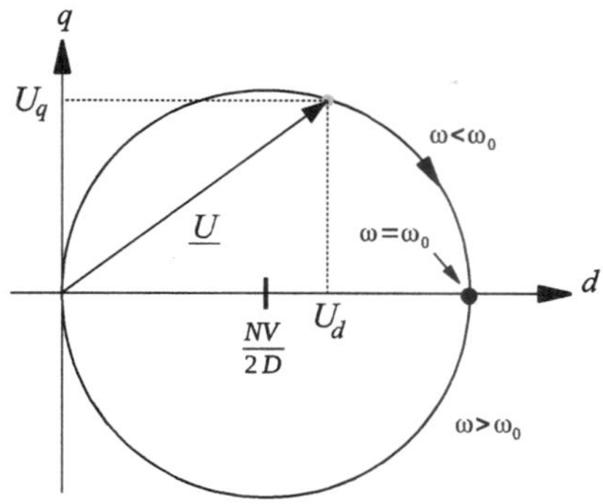

Figure 3.9 Evolution of the extremity of vector $\underline{U}$ in the rotating reference frame when ω is varying; $V_q = 0$, V_d constant.

The ratio $\frac{\underline{f}_r}{\underline{\dot{w}}}$ is named the *acoustic impedance* of the load. Eq. (3.27) yields

$$\underline{F}_r = \frac{\tilde{K} - \tilde{M}\omega^2}{j\omega}\underline{U} + \tilde{D}\underline{U}. \tag{3.28}$$

By splitting Eq. (3.28) into its real and imaginary part, we obtain

$$F_{rd} = \tilde{D}U_d + \frac{\tilde{K} - \tilde{M}\omega^2}{j\omega}U_q \quad \text{and} \quad F_{rq} = \tilde{D}U_q - \frac{\tilde{K} - \tilde{M}\omega^2}{j\omega}U_d. \tag{3.29}$$

By substituting Eq. (3.29) into (3.22), we obtain two new equations:

$$\begin{aligned} \frac{(K+\tilde{K}) - (M+\tilde{M})\omega^2}{\omega}U_q + (D+\tilde{D})U_d = NV, \\ -\frac{(K+\tilde{K}) - (M+\tilde{M})\omega^2}{\omega}U_d + (D+\tilde{D})U_q = 0. \end{aligned} \tag{3.30}$$

As it can be seen, Eq. (3.30) is equivalent to Eq. (3.23), if K is replaced by $K+\tilde{K}$, M by $M+\tilde{M}$, and D by $D+\tilde{D}$. Therefore, the load has an effect on the evolution of the vector $\underline{U}$ when ω is varying: it remains a circle, but with a modified radius $\frac{NV}{2(D+\tilde{D})}$. This effect also changes the resonance frequency to $\tilde{\omega}_0 = \sqrt{\frac{K+\tilde{K}}{M+\tilde{M}}}$. However, condition $U_q = 0$ still corresponds to the resonance.

3.3.2 Dynamic modeling in the rotating reference frame

In this section, the derivatives of W_d, W_q, U_d, U_q are not constant. The first derivative of w gives

$$\dot{\underline{w}} = \left(\left(\dot{W}_d + j\dot{W}_q\right) + j\omega\left(W_d + jW_q\right)\right)e^{j\omega t} = (U_d + jU_q)e^{j\omega t}. \quad (3.31)$$

By identifying the two sides of Eq. (3.31), we can write

$$U_d = \dot{W}_d - \omega W_q \quad \text{and} \quad U_q = \dot{W}_q + \omega W_d. \quad (3.32)$$

It is then possible to calculate $\ddot{w}$ as

$$\ddot{w} = \left(\left(\dot{U}_d - \omega U_q\right) + j\left(\dot{U}_q + \omega U_d\right)\right)e^{j\omega t}. \quad (3.33)$$

Using Eq. (3.32) leads to

$$W_d = \frac{U_q}{\omega} + \frac{\dot{U}_d}{\omega^2} - \frac{\ddot{W}_d}{\omega^2}. \quad (3.34)$$

If we take all the terms of Eq. (3.34) into account, the model will be precise, but complex. This is why we introduce here the *dynamic assumption* in order to reduce the model's complexity, without loss of practicability. Indeed, in most cases, the dynamics of W_d is limited by ω; it means that the bandwidth of W_d is lower than the resonance pulsation of the vibrating system. This condition is valid for high frequency vibrating mode, and leads to neglecting $\frac{\ddot{W}_d}{\omega^2}$ beyond W_d for a nonzero amplitude vibration. This assumption is supposed to be valid in our models, provided that the vibration's response time is not too low.

According to the *dynamic assumption*, and following a similar process with W_q, one can write

$$\underline{w} = \left(\left(\frac{U_q}{\omega} + \frac{\dot{U}_d}{\omega^2}\right) + j\left(-\frac{U_d}{\omega} + \frac{\dot{U}_q}{\omega^2}\right)\right)e^{j\omega t}. \quad (3.35)$$

Substituting (3.35), (3.33) and (3.31) into (3.1), and identifying the real and imaginary parts, we obtain two new equations, describing the dynamic of the vibration velocity in the rotating reference frame:

$$\begin{aligned} \left(M + \frac{K}{\omega^2}\right)\dot{U}_d + DU_d + \left(\frac{K}{\omega} - M\omega\right)U_q &= F_{pd} - F_{rd}, \\ \left(M + \frac{K}{\omega^2}\right)\dot{U}_q + DU_q - \left(\frac{K}{\omega} - M\omega\right)U_d &= F_{pq} - F_{rq}. \end{aligned} \quad (3.36)$$

The *dynamic assumption*

The *dynamic assumption* is very helpful in simplifying the model in the (d, q) frame. But how can we know if this hypothesis is valid for a system? To address this question, we introduce $\widehat{W}_d(j\Omega)$, the Fourier transform of $W_d(t)$. We consider that the dynamics of W_d is limited by Ω_M, which means:

$$\widehat{W}_d(j\Omega) = 0 \quad \text{if} \quad \Omega > \Omega_M \quad \text{or} \quad \Omega < -\Omega_M. \tag{3.37}$$

The Fourier transform is invertible, and W_d is continuous function of time; therefore, we have

$$W_d(t) = \int_{-\infty}^{\infty} \widehat{W}_d e^{jt\Omega}\,\mathrm{d}\Omega = \int_{-\Omega_M}^{\Omega_M} \widehat{W}_d e^{jt\Omega}\,\mathrm{d}\Omega. \tag{3.38}$$

Due to the differentiation properties of $\widehat{W}_d(j\Omega)$, the Fourier transform of the second derivative of W_d can be calculated as

$$\widehat{\ddot{W}}_d = -\Omega^2 \widehat{W}_d, \tag{3.39}$$

which equals 0 if $\Omega > \Omega_M$ or $\Omega < -\Omega_M$. Again, by using the inverse Fourier transform, we write:

$$\begin{aligned}\left|\frac{\ddot{W}_d(t)}{\omega^2}\right| &= \left|\int_{-\infty}^{\infty} \frac{\Omega^2}{\omega^2} \widehat{W}_d e^{jt\Omega}\,\mathrm{d}\Omega\right| = \left|\int_{-\Omega_M}^{\Omega_M} \frac{\Omega^2}{\omega^2} \widehat{W}_d e^{jt\Omega}\,\mathrm{d}\Omega\right| \\ &\le \left|\int_{-\Omega_M}^{\Omega_M} \frac{\Omega_M^2}{\omega^2} \widehat{W}_d e^{jt\Omega}\,\mathrm{d}\Omega\right| = \frac{\Omega_M^2}{\omega^2}\left|\int_{-\Omega_M}^{\Omega_M} \widehat{W}_d e^{jt\Omega}\,\mathrm{d}\Omega\right| \\ &\le \frac{\Omega_M^2}{\omega^2}\left|W_d(t)\right|.\end{aligned} \tag{3.40}$$

It is evident from Eq. (3.40) that the *dynamic assumption* can be ensured if $\frac{\Omega_M^2}{\omega^2} \ll 1$: the bandwidth of W_d, which is related to its dynamics, has to be much smaller than the working pulsation of the piezoelectric transducer. In a second-order system, the ratio $\frac{\omega^2}{\Omega_M^2}$ is called the quality factor. Hence, the model is valid only for high quality factor transducers. Fortunately, transducers with a quality factor of 100 or more are common. This is why, in open-loop, the *dynamic assumption* is supposed to be always valid for the systems presented in this book. However, one objective of the closed-loop control is to make the system faster, increasing the bandwidth of the system. Therefore to preserve the *dynamic assumption* valid, we cannot achieve a closed-loop control that would respond in less than 10 periods of the vibration itself with the vector control presented here.

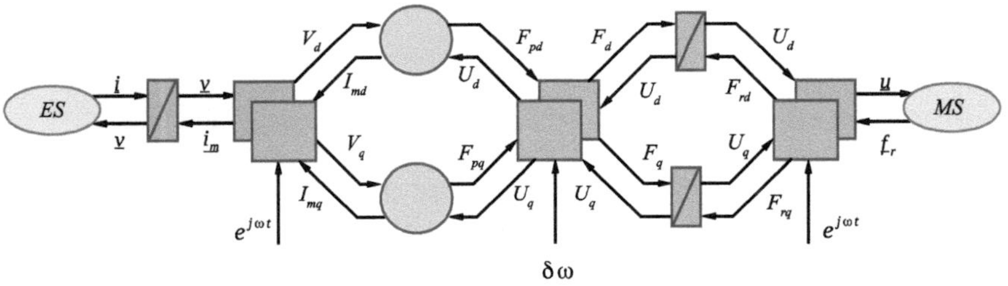

Figure 3.10 Energetic macroscopic representation of the system in the rotating reference frame.

These two new equations constitute the model of the piezoelectric transducer in the rotating reference frame. They show that the forces in axis d control the dynamics of the vibration velocity U_d, the environment acts as a perturbation (through F_{rd}) and there is a coupling with the axis q. The same conclusion holds for axis q which is controlled by F_q with perturbation and coupling from the other axis. In order to simplify the model, we introduce a simplification, based on the fact that piezoelectric actuators generally have a high quality factor. It is therefore admitted that the pulsation of the vibration is close to the pulsation of resonance ω_0. We denote by $\delta\omega$ the deviation of ω from ω_0 such that $\omega = \omega_0 + \delta\omega$ with $\delta\omega \ll \omega_0$. We consider then that $M + \frac{K}{\omega^2} \simeq 2M$ and $\frac{K}{\omega} - M\omega \simeq -2M\delta\omega$. Equations in (3.36) simplify to

$$\begin{aligned} 2M\dot{U}_d + DU_d - 2M\delta\omega U_q &= F_{pd} - F_{rd}, \\ 2M\dot{U}_q + DU_q + 2M\delta\omega U_d &= F_{pq} - F_{rq}. \end{aligned} \tag{3.41}$$

We introduce two forces as follows:

$$F_d = F_{pd} + 2M\delta\omega U_q \quad \text{and} \quad F_q = F_{pq} - 2M\delta\omega U_d. \tag{3.42}$$

And we finally write:

$$\begin{aligned} 2M\dot{U}_d + DU_d &= F_d - F_{rd}, \\ 2M\dot{U}_q + DU_q &= F_q - F_{rq}. \end{aligned} \tag{3.43}$$

Eqs. (3.43) are now used to represent the piezoelectric transducer by using the EMR, as depicted in Fig. 3.10. In this new representation, we use the complex forms of v, i, u and f_r, since these variables are supposed to be sinusoidal functions of time.

Table 3.2 Simulink blocks for the 3 new couplings.

$(V_d + jV_q) = \underline{v}e^{-j\omega t}$ $\underline{i}_m = (I_{md} + jI_{mq})\, e^{j\omega t}$	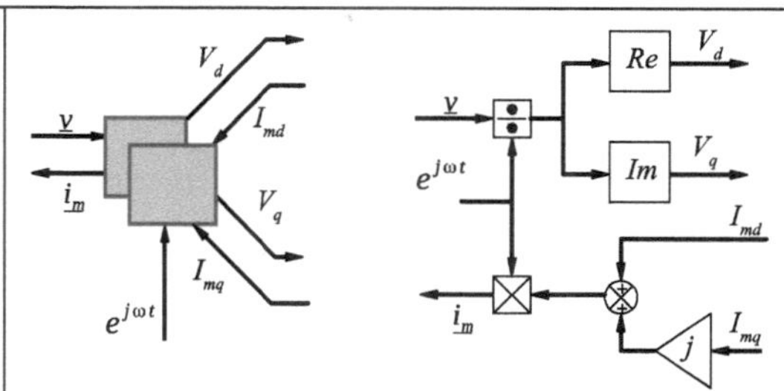
$(F_{rd} + jF_{rq}) = \underline{f}_r e^{-j\omega t}$ $\underline{u} = (U_d + jU_q)\, e^{j\omega t}$	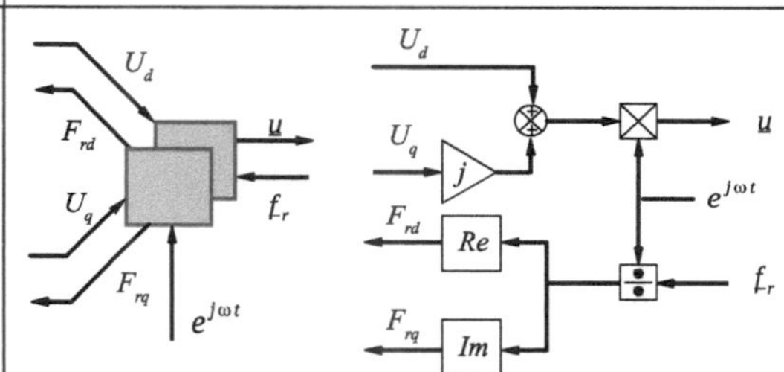
$F_d = F_{pd} + 2M\delta\omega U_q$ $F_q = F_{pq} - 2M\delta\omega U_d$	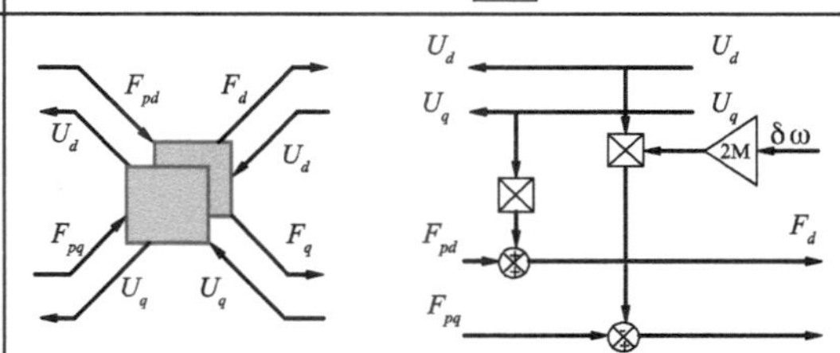

This EMR clearly shows that the dynamics of the two axes are coupled, but this coupling depends on $\delta\omega$, the difference between the working pulsation and that of the resonance. This coupling is very similar to the coupling that exists in electromagnetic machines. In these machines too, the coupling depends on the working pulsation of the voltage and current. However, with piezoelectric actuators, the coupling is hardly predictable, mostly because the resonance frequency strongly varies with the operating conditions.

Since Fig. 3.10 introduces three new couplings, we detail in Table 3.2 how these couplings are converted into Simulink blocks.

3.3.3 Simulation results

The model is implemented into Matlab–Simulink using the same parameters as in Sect. 3.1.4. In the first try, we maintain $V_q = 0$ and vary V_d from 0 to 100 V with a square waveform at a frequency of 2 Hz. At $t = 0$, we

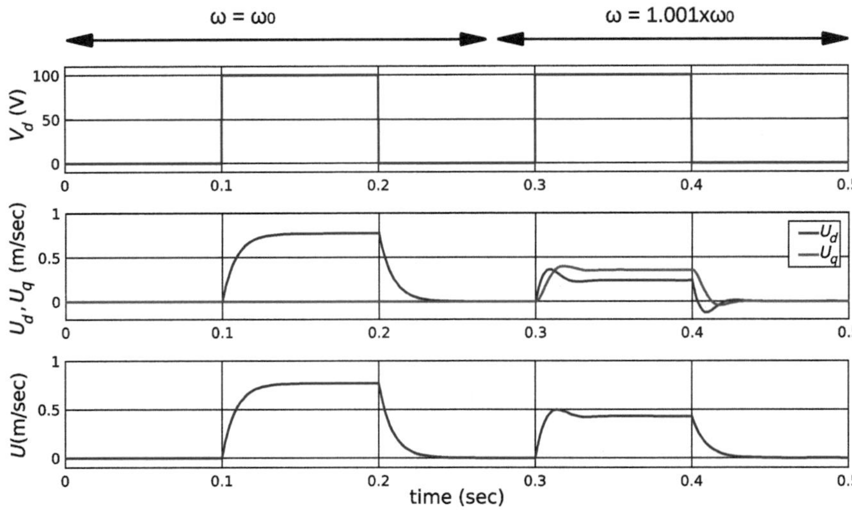

Figure 3.11 Simulation of the transitory response of a Langevin transducer.

consider that the transducer is operated at resonance and $\omega = \omega_0$, leading to $\delta\omega = 0$. Then, at $t = 250$ ms, the operating pulsation is set to $\omega = 1.001\omega_0$, leading to $\delta\omega = 0.001\omega_0$. The simulation results are presented in Fig. 3.11.

When $\delta\omega = 0$, there is no coupling between axes d and q. Since $V_q = 0$, U_q remains equal to 0, while U_d follows a first-order type response, as predicted by Eq. (3.43). When $\delta\omega \neq 0$, the coupling between the two axes is not null. The voltage in the d axis makes U_d increase. This, in turn, creates a coupling term in axis q, which makes U_q increase. Consequently, in steady state U_q is not null, and U_d is lower than its value at resonance. Moreover, due to the coupling, the evolutions of U_d and U_q during the transitory period are not of the first-order type, but show overshoot, which is typical of a second-order type system.

3.4. Modeling a traveling wave in a linear beam

We have presented in the previous section a model of a Langevin transducer in the rotating reference frame. This model directly predicts the evolution of the vibration amplitude when the excitation's pulsation is close to that of the resonance. This model can be extended to the cases of the piezoelectric transducers using only one vibration mode.

In other cases, the piezoelectric actuators use several vibration modes, in particular those exploiting a traveling wave. In this section, we apply the same method to such systems.

3.4.1 The multimodal approach

Two theories can be used to explain how two exciters placed on a beam can propagate a traveling wave. The first theory considers one exciter as a mechanical source and the other as a damper. If the beam had an infinite length, the exciter would create a traveling wave, denote by $w_+(x, t)$, at any vibration frequency. An example of a traveling wave is given by

$$w_+(x, t) = W \sin(\omega t - kx). \tag{3.44}$$

We remark that the nodes of vibration (i.e., where $w_+ = 0$) are obtained at position equal to $x = \frac{\omega t + n\pi}{k}$ where n is an integer. Therefore, the vibration nodes are not at constant position, but depend on time, which is typical of a traveling wave.

With a finite length, the wave reflects on the edges, leading to an inverse traveling wave, which propagates with the same amplitude, but in the reverse direction, named $w_-(x, t)$. We write then

$$w_-(x, t) = -W \sin(\omega t + kx + \varphi). \tag{3.45}$$

The phase shift φ has been introduced in Eq. (3.45) in order to take into account the propagation time between the exciter and edge. Therefore, the resulting wave is the sum of w_+ and w_-, and by adequately choosing the reference time, we obtain

$$w(x, t) = 2 \sin(\omega t) \cos(kx - \varphi/2). \tag{3.46}$$

In Eq. (3.46), the vibration nodes are located at constant positions $x = \frac{\frac{\pi - \varphi}{2} + n\pi}{k}$, which is typical of a standing wave.

Based on this theory, we can prevent the reflection of the wave, and thus obtain a pure traveling wave, if a second exciter is added to absorb it. This can be actually achieved experimentally, provided that a matching impedance is connected to the electric terminals of this second exciter, and if it is precisely placed on the beam. This is illustrated by Fig. 3.12.

In another theory, a traveling wave is obtained in a finite beam if its vibrating modes are energized synchronously, at the same frequency, with a specific pattern of excitation. This is the *multimodal approach*. In theory,

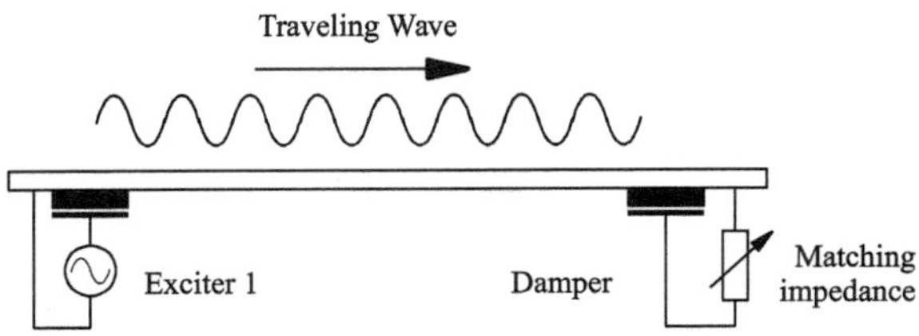

Figure 3.12 A traveling wave generated by an exciter and absorbed by a damper to avoid reflection.

the excitation conditions are very difficult to fulfill, and need a lot of exciters. However, to simplify the problem, it has been practically shown that a traveling wave can be obtained if two successive modes are energized at the same frequency, but shifted in time by 90°. However, since each mode has its resonant frequency, it is necessary to vibrate the plate at a frequency between the two modes' resonance frequency to not favor one mode over the other. With this *multimodal approach*, the excitation of the beam must ensure three conditions for obtaining the progressive wave: two amplitude conditions and a phase condition. With only one exciter, only two conditions can be tuned, like, for instance, the amplitude of one mode and phase, by adjusting the voltage on the exciter and the frequency. But to achieve a pure traveling wave, a second exciter is needed to add more control inputs.

In this part, we follow the theory of the *multimodal* approach, since it is more general. Therefore, we will build a model which predicts the vibration displacement of each mode. This model will be used for a control aiming at synchronizing the modes to obtain a traveling wave.

3.4.2 Model of a beam with two exciters

We consider a beam of length L actuated by two exciters. Index 1 is for the first exciter, and 2 indicates the second. The two exciters are denoted E_1 and E_2, and they are placed at the positions x_1 and x_2 on the beam as detailed in Fig. 3.13. As presented in Chap. 2, the deformation of the beam is the sum of the contributions of an infinite number of modes, and we write

$$w(x,t) = \sum \Phi_n(x) w_n(t), \tag{3.47}$$

where $\underline{w}$ is the deformation all along the beam, $\underline{w}_n$ is the displacement of the mode n and $\Phi_n(x)$ is the deformation mode shape of the mode n. In this model, we consider only two modes, denoted by the modes a and b, which are the two successive modes energized to obtain the traveling wave. Therefore, Eq. (3.47) simplifies to

$$w(x, t) = \Phi_a(x)w_a(t) + \Phi_b(x)w_b(t). \tag{3.48}$$

Chapter 2 gives the dynamics of each mode. They follow a second-order type equation, and we write:

$$\begin{aligned} M_a\ddot{w}_a + D_a\dot{w}_a + K_a w &= f_a, \\ M_b\ddot{w}_b + D_b\dot{w}_b + K_b w &= f_b, \end{aligned} \tag{3.49}$$

where M_i, D_i, K_i, $i \in \{a, b\}$, are the modal mass, damping and stiffness for the modes a and b, and F_a and F_b are the forces exerted by the exciters on the beam for each mode.

The two exciters contribute to the forces F_a and F_b but not equally. Indeed, the contributions depend on their relative position on the beam. To be concise, and without loss of generality, the two forces F_a and F_b can be written as:

$$\begin{aligned} f_a &= N\left(\Phi_a(x_1)v_1 + \Phi_a(x_2)v_2\right), \\ f_b &= N\left(\Phi_b(x_1)v_1 + \Phi_b(x_2)v_2\right), \end{aligned} \tag{3.50}$$

where N is the electromechanical force factor. If we introduce the matrix Φ_{12} given by

$$\Phi_{12} = \begin{pmatrix} \Phi_a(x_1) & \Phi_a(x_2) \\ \Phi_b(x_1) & \Phi_b(x_2) \end{pmatrix}, \tag{3.51}$$

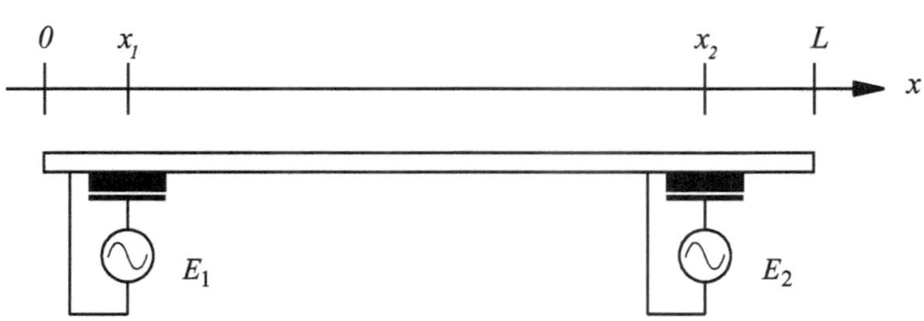

Figure 3.13 A beam and two exciters.

How can we represent a traveling wave in a textbook?

If we consider a pure traveling wave such as $w(x, t) = W\sin(\omega t - kx)$ then we can draw the deformation of the beam at different time steps, as shown in Fig. 3.14. In this figure, we also represented the vibration nodes for each time step. The positions of the vibration nodes on the beam are not fixed, they vary with time, which is typical of a traveling wave.

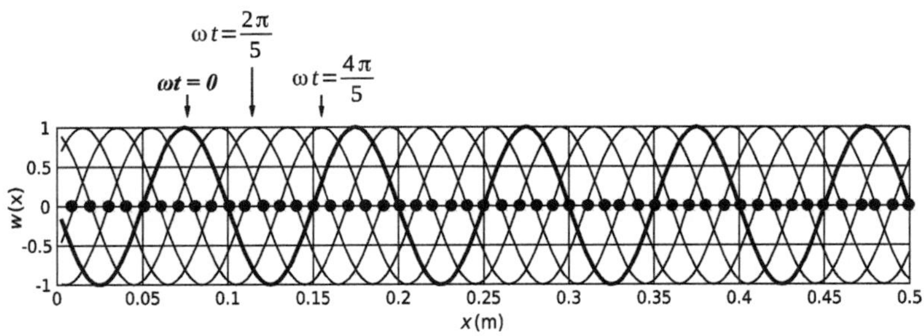

Figure 3.14 Representation of a traveling wave at different time steps; the • represent the vibration nodes.

For sake of clarity, only few time steps can be represented, but usually, it's enough to have a qualitative evaluation of the traveling wave. For a more advanced analysis, we use a 2D plot, in a complex plane. Indeed, we can also write:

$$\begin{aligned} w(x, t) &= W\sin(\omega t)\cos(kx) - W\cos(\omega t)\sin(kx) \\ &= W_R\sin(\omega t) - W_I\sin(\omega t) \end{aligned} \tag{3.52}$$

where $W_R = W\cos(kx)$ and $W_I = W\sin(kx)$. In a real-imaginary frame, the points $M(W_R; W_I)$ are along a circle, as depicted Fig. 3.15.

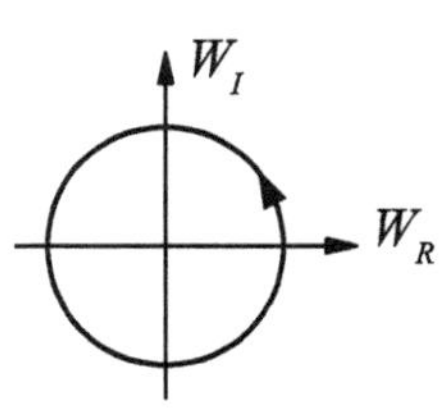

Figure 3.15 Representation of a traveling wave in a real-imaginary frame.

This representation is more abstract, but has the main advantage to be concise: a perfect traveling wave will always be represented as a circle, and all the representations not showing a perfect circle are typical of a not perfect traveling wave.

then Eq. (3.50) can be written in the matrix form, using $f_{p1} = Nv_1$ and $f_{p2} = Nv_2$, the two forces produced by the exciters, as

$$\begin{pmatrix} f_a \\ f_b \end{pmatrix} = \Phi_{12} \begin{pmatrix} f_{1p} \\ f_{2p} \end{pmatrix}. \tag{3.53}$$

In the same way, from Eq. (3.48), it is obvious that the matrix form can also be used for the displacement and velocity amplitude of each mode and leads to

$$\begin{pmatrix} u_1 \\ u_2 \end{pmatrix} = \Phi_{12} \begin{pmatrix} u_a \\ u_b \end{pmatrix} \tag{3.54}$$

where u_1 and u_2 are the vibration velocities at positions x_1 and x_2; u_a and u_b are the modal vibration velocities.

It is now possible to model the system in the rotating reference frame. Following the notations of Section 3.3.2, we write

$$\begin{aligned} 2M_i\dot{U}_{id} + DU_{id} &= F_{id}, \\ 2M_i\dot{U}_{iq} + DU_{id} &= F_{iq}, \end{aligned} \tag{3.55}$$

as well as

$$F_{id} = F_{ipd} + 2M_i\delta\omega_i U_{iq} \quad \text{and} \quad F_{iq} = F_{ipq} - 2M_i\delta\omega_i U_{id} \tag{3.56}$$

with $i \in \{a, b\}$. Moreover, Eqs. (3.53) and (3.56) yield:

$$\begin{aligned} F_{ipd} &= \Phi_i(x_1)F_{1d} + \Phi_i(x_2)F_{2d}, \\ F_{ipq} &= \Phi_i(x_1)F_{1q} + \Phi_i(x_2)F_{2q}. \end{aligned} \tag{3.57}$$

The energetic macroscopic representation in the rotating reference is then depicted in Fig. 3.16. It is composed of two second-order systems which operate independently, but which are actually coupled at the level of the excitation forces. It should be emphasized here that the coupling between the axes d and q for mode a depends on $\delta\omega_a$ which represents the difference between the resonance pulsation of the mode a and the pulsation of the supply voltages. Therefore, this coupling differs from the coupling for mode b since $\omega_{0a} \neq \omega_{0b}$. In this representation, the external forces have been represented with external mechanical sources (MS), even if they are considered null in this part.

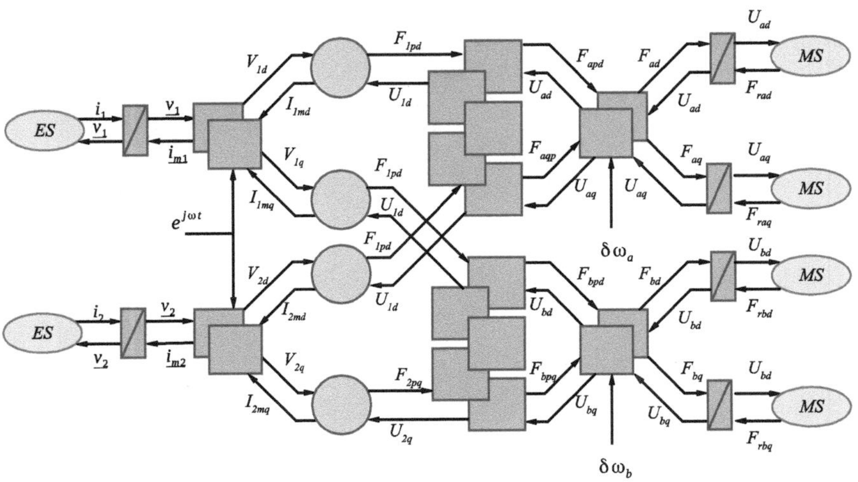

Figure 3.16 EMR of a beam with two exciters, named 1 and 2, energizing two modes, denoted *a* and *b*.

3.4.3 Simulation results

We consider a beam of length $L = 493$ mm, with a square section of 6×6 mm^2, made of aluminum. Two exciters are located at $x_1 = 47$ mm and $x_2 = 446$ mm. The modal analysis of this beam is presented in Fig. 3.17. We select two modes: mode a is at 19440 Hz, mode b is at 21490 Hz. Their deformation mode shapes, $\Phi_a(x)$ and $\Phi_b(x)$, are presented in Fig. 3.18.

The modal analysis made it possible to calculate the parameters of the equivalent model for each mode. They are given in Table 3.3. Moreover, the force factor equals 0.1 N/V for the 2 transducers.

Fig. 3.19 shows the transitory response of the vibration amplitude, in the rotating reference frame, for a step variation of the two voltages v_1 and v_2.

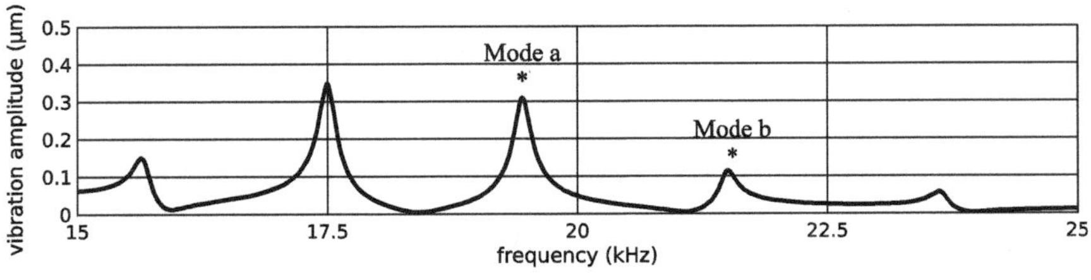

Figure 3.17 Modal analysis of the beam, showing the vibration amplitude of a point located at $x = 123$ mm, as a function of frequency for a constant voltage amplitude (corresponding to 1 N on E_1 and 0 N for E_2).

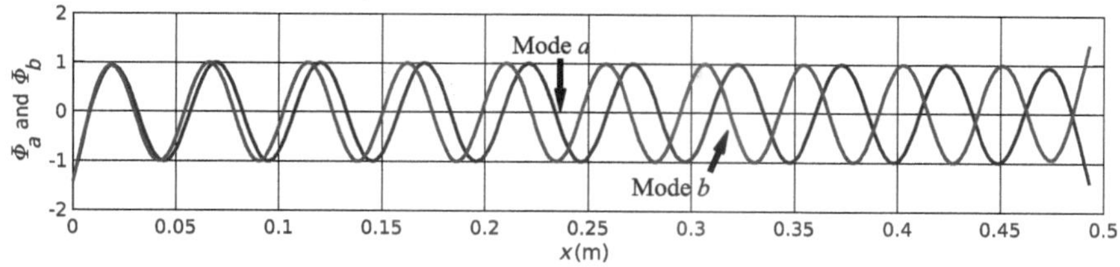

Figure 3.18 Deformation mode shapes for modes *a* and *b* as a function of *x*.

Table 3.3 Beam simulation parameters, derived from modal analysis.

Mode	K_i	M_i	D_i	$\Phi_i(x_1)$	$\Phi_i(x_2)$
a	336 N/μm	23 g	23 Ns/m	−0.9311	−0.9451
b	410 N/μm	23 g	23 Ns/m	−0.7777	0.8105

The pulsation has been set in the middle of the two modes. The transient time is approximately equal to 10 ms, and the response is very oscillating, because the operating pulsation of the system is far from the resonance of the two modes: $\delta\omega_a$ and $\delta\omega_b$ are high, so are the coupling terms between the axes *d* and *q*. In steady state, we obtain $U_{da} = 0$, $U_{qb} = 0$ and $U_{qa} = U_{db}$. This can also be written as $\underline{U}_a = j\underline{U}_b$, which is the necessary condition to obtain a traveling wave from the *multimodal approach*.

In order to look more closely at the steady state, we draw in Fig. 3.20 the voltages v_1 and v_2, the vibration velocity amplitude of the two modes $\underline{U}_a$ and $\underline{U}_b$, and the corresponding representation of the traveling wave for three pulsation conditions:

Case 1. $\omega = \frac{\omega_a+\omega_b}{2}$;

Case 2. $\omega = \frac{\omega_a+\omega_b}{2} + 1600$ rad/s with the same voltage as in Case 1;

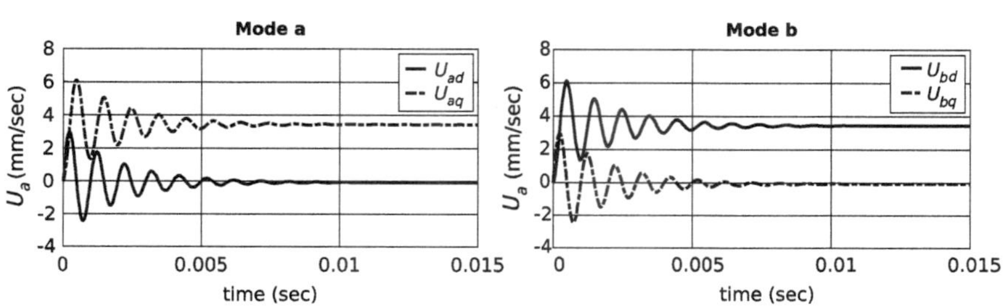

Figure 3.19 Transitory response of the vibration velocity amplitudes for each mode, and in the rotating reference frame, for a step variation of the voltages; the operating pulsation is set in the middle of the two modes.

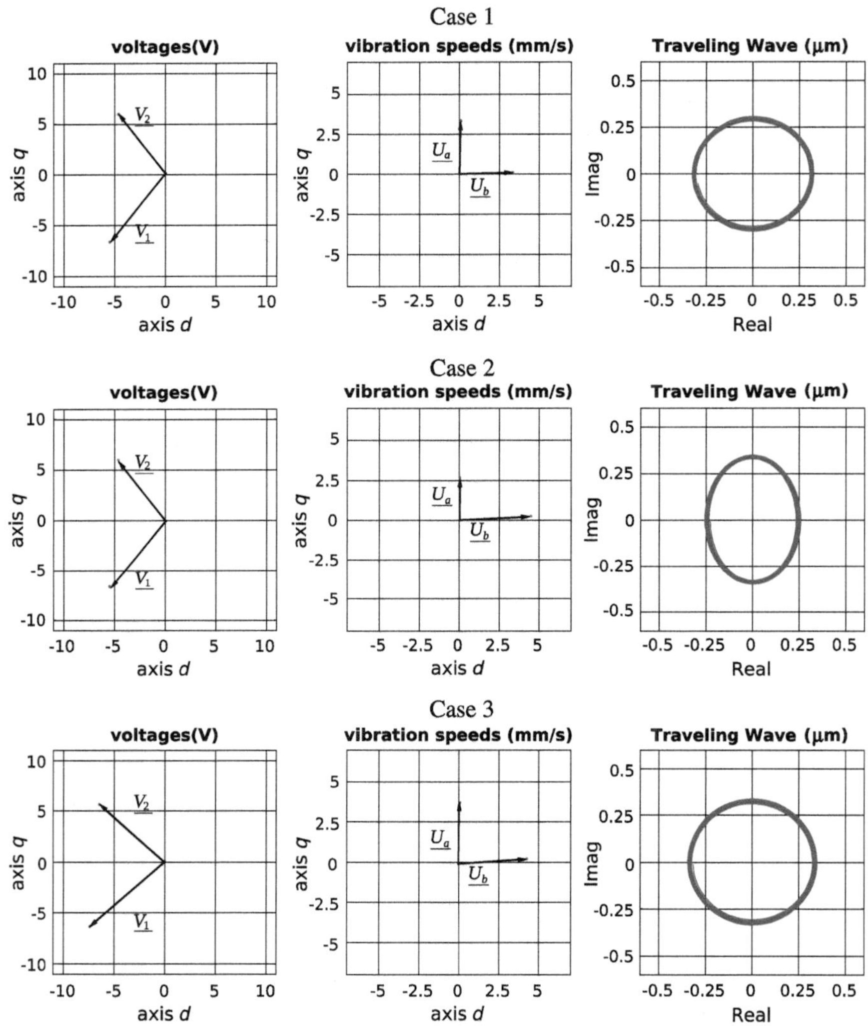

Figure 3.20 Simulation results of the modal vibration velocities and the corresponding Traveling wave for given voltages; results are presented in the rotating reference frame and the real-imaginary plane for the traveling wave. Case 1: $\omega = \frac{\omega_a+\omega_b}{2}$; case 2: $\omega = \frac{\omega_a+\omega_b}{2} + 1600$ rad/s with the same voltages as in case 1; case 3: $\omega = \frac{\omega_a+\omega_b}{2} + 1600$ rad/s, with optimized voltages in order to have $\underline{U}_a = j\underline{U}_b$.

Case 3. $\omega = \frac{\omega_a+\omega_b}{2} + 100$ rad/s, with adapted voltages in order to have mode $\underline{U}_a = j\underline{U}_b$.

It can be seen from the results of Case 1 that due to the positions of E_1 and E_2, the voltages applied to the exciters don't have equal amplitude and they are out of phase by $\frac{\pi}{2}$, but the velocities are pretty much the same. This actually leads to a traveling wave, with a representation which almost follows a circle.

For Case 2, pulsation ω is increased while maintaining the same voltages: the vibration velocities' amplitude $\underline{U}_b$ increases, because we operate closer to mode b. Also, the phase difference between $\underline{U}_a$ and $\underline{U}_b$ is not $\frac{\pi}{2}$, therefore the condition to obtain a perfect traveling wave is not fulfilled, which is confirmed by the observation of the representation of the traveling wave: the locus of the vibration of the beam is not along a circle, but along an ellipse.

To again have $\underline{U}_a = j\underline{U}_b$, we have to adjust the voltages applied to the exciters, which is done in Case 3.

These simulations show that, in spite the fact that all the modes are involved in the traveling wave, condition $\underline{U}_a = \pm j\underline{U}_b$ is sufficient to have a traveling wave. It also highlights the fact that due to the couplings inside the system (coupling between axes d and q, coupling between modes a and b) this condition does not necessarily lead to supplying two voltages in quadrature for the exciters, a condition which is generally admitted. However, tuning the voltage is very difficult; this is why, in the next section, we will show how to automatically tune the voltages by a closed-loop control to obtain a traveling wave, which is robust besides the operating conditions, like the pulsation ω, for instance.

3.4.4 Modeling of a Traveling Wave Ultrasonic Motor in the rotating reference frame

Traveling Wave Ultrasonic Motors (TWUM) exploit two vibration modes of a stator to propel by friction a rotor pressed on it. For that, an elliptic motion of the stator's particle is required. This is achieved using thick stator, because we will see that the tangential motion of the particles and the normal one are linked in a vibrating stator. If the stator has an annular shape, then the two successive modes chosen to create the traveling wave can have the same resonance frequency, but also the same characteristics, which really simplifies the design of TWUM. In the following section, the basic design rules of a TWUM are used on order to adapt the model of the beam with a traveling wave. Then, this model is simplified within the framework of specific assumptions.

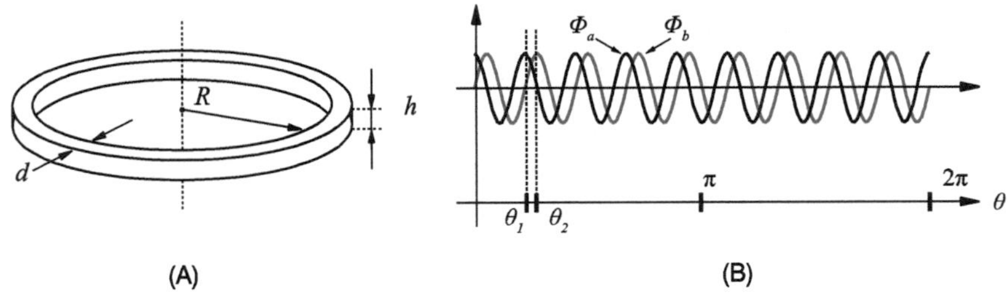

Figure 3.21 Design of the stator (A) and its deformation mode shapes (B).

3.4.4.1 The model with two independent phases

The stator is an annulus (Fig. 3.21), with radius denoted by R and having thickness h; the width of the annulus, d, is neglected with respect to R for sake of conciseness. Due to the geometry of the stator, which is invariant to any rotation around its axis, it is possible to find two vibration modes a and b such that

$$\Phi_a = \cos(k\theta) \quad \text{and} \quad \Phi_b = \sin(k\theta), \tag{3.58}$$

where k is an integer, and θ is the angular position on the annulus. The relations (2.130)–(2.134) on page 92, give then the same values for M_i, K_i and D_i, $i \in \{a, b\}$.

Two independent exciters E_1 and E_2 are positioned on the annulus in order to excite these two modes of vibration; the positions of E_1 and E_2 are denoted θ_1 and θ_2, respectively. We obtain then a system similar to the system studied in Sect. 3.1.4, but with a circular organization, and the model depicted Fig. 3.16 can be used again. The matrix Φ_{ab} is then equal to

$$\Phi_{ab} = \begin{pmatrix} \cos(\theta_1) & \cos(\theta_2) \\ \sin(\theta_1) & \sin(\theta_2) \end{pmatrix}. \tag{3.59}$$

Hence, the model of a TWUM describes the 4 mechanical sources depicted in Fig. 3.16, and should therefore give a model for F_{rad}, F_{rbd}, F_{raq} and F_{rbq}. Also, it should describe how torque is produced from the vibration velocities U_{ad}, U_{aq}, U_{bd} and U_{bq}. The modeling of the TWUM is then split into four parts:

- The kinematics of the stator, which describes the relationship between the vibration velocity of the stator and the rotation and translation velocities of the rotor;

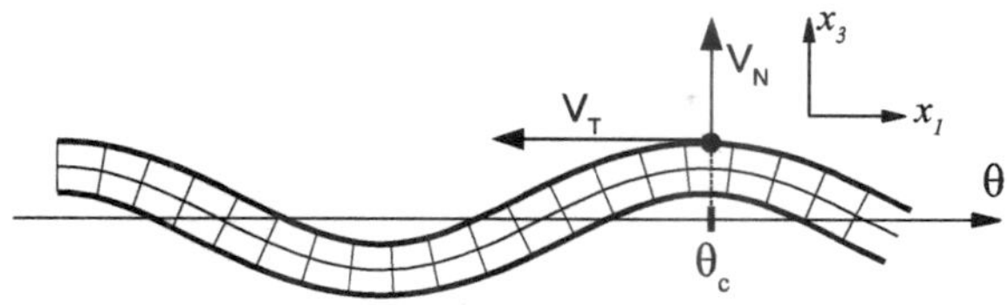

Figure 3.22 Deformation of the stator; the deformations along the normal axis (x_3) and the tangential axis (x_1) are exaggerated.

- The equilibrium of the stator, which projects the forces that the rotor exerts on the stator on the vibration modes;
- The dynamic of the rotor, which describes the evolution of the rotation speed;
- The torque–speed characteristics.

Kinematic study of the stator. Fig. 3.22 depicts a closed loop of the stator's deformation.

If we denote by w_a and w_b the deformation for modes a and b, Eq. (3.58) leads to

$$w_3 = w_a \cos(k\theta) + w_b \sin(k\theta). \tag{3.60}$$

In fact, when the stator undergoes a deformation in the normal axis, the particles are also deformed in the tangential axis, and the theory of thin plates leads to:

$$w_1 = -\frac{h}{2R}\frac{\partial w_3}{\partial \theta} = k\frac{h}{2R} w_a \sin(k\theta) - k\frac{h}{2R} u_b \cos(k\theta), \tag{3.61}$$

where w_1 is the deformation along the tangential axis of the particles at the top of the stator, u_a and u_b are the vibration velocities for the modes. If a rotor is mounted on the stator, they become in contact. The study of the contact conditions is very difficult, and very important for the design of TWUM (Wallaschek, 1998). This level of complexity and precision is not mandatory in this model for the control. Therefore, we make the assumption that the contact between the stator and rotor is *punctual.* This corresponds to idealized contact conditions. We call them the *ideal rotor* contact conditions. The position θ_c at which the contact occurs is given by the maximum of the displacement along x_3. Indeed, from Eq. (3.60), we obtain

$$k\theta_c = a\tan\left(\frac{w_b}{w_a}\right). \tag{3.62}$$

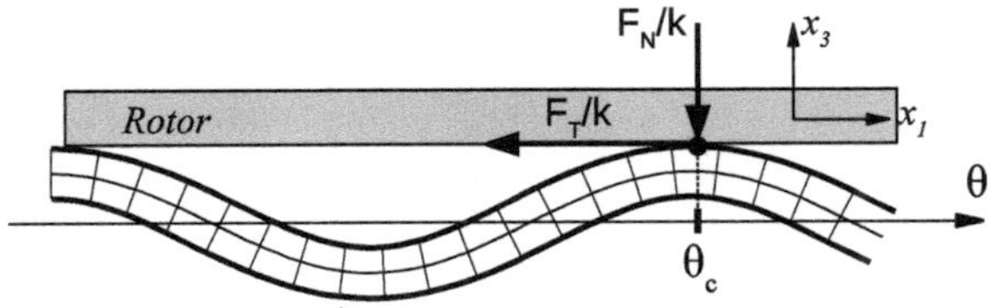

Figure 3.23 Normal (F_N) and tangential (F_T) forces applied on the stator at a contact point.

The rotor of a TWUM has 2 degrees of freedom:

- The rotation about x_3, and we denote Ω_{id} the rotational speed,
- The translation along x_3 is equal to the vibration amplitude, and we denote V_{Nid} the normal speed.

At the contact point, the velocity of the stator equals that of the rotor, and if we introduce $V_{Tid} = R\Omega_{id}$, the tangential speed, differentiating Eqs. (3.60) and (3.61) produces

$$V_{Nid} = u_a \cos(k\theta_c) + u_b \sin(k\theta_c) \tag{3.63}$$

and

$$V_{Tid} = k\frac{h}{2R}u_a \sin(k\theta_c) - k\frac{h}{2R}u_b \cos(k\theta_c). \tag{3.64}$$

For sake of clarity, we introduce here V'_{Tid} such that $V_{Tid} = -k\frac{h}{2R}$. It is then possible to write

$$\left(V_{Nid} + jV'_{Tid}\right) = e^{-jk\theta_c}\left(u_a + ju_b\right) = e^{-jk\theta_c}\underline{u}, \tag{3.65}$$

the projection of the contact forces onto the vibration modes of the stator: the contact forces F_N and F_T along the normal (x_3) and tangential (x_1) axes are depicted in Fig. 3.23.

They are punctual, and located at the k contact points. Therefore, the stress they produce can be written as

$$t = \frac{F_N}{k}\delta(\theta - i\theta_c)x_3 + \frac{F_T}{k}\delta(\theta - i\theta_c)x_1, \quad i \in [1, k]. \tag{3.66}$$

The projection of t on the vibration mode a gives f_{ra} as

$$\begin{aligned} f_{ra} &= \int_0^{2\pi} \left(\frac{F_N}{k}\delta(\theta - i\theta_c)\cos(k\theta) + \frac{F_T}{k}\delta(\theta - i\theta_c)\frac{-kh}{2R}\sin(k\theta)\right) d\theta \\ &= F_N \cos(k\theta_c) - k\frac{h}{2R}F_T \sin(k\theta_c). \end{aligned} \tag{3.67}$$

By definition, the vector $\underline{f}_{ra}$ is defined such that $f_{ra} = \Im\left(\underline{f}_{ra}\right)$. We can then use Eq. (3.67) to write

$$\underline{f}_{ra} = j\left(F_N + jk\frac{h}{2R}F_T\right)e^{jk\theta_c}. \tag{3.68}$$

In the same way, we can calculate f_{rb} as

$$\begin{aligned} f_{rb} &= \int_0^{2\pi}\left(\frac{F_N}{k}\delta(\theta - i\theta_c)\sin(k\theta) + \frac{F_T}{k}\delta(\theta - i\theta_c)k\frac{h}{2R}\cos(k\theta)\right)\mathrm{d}\theta \\ &= F_N\sin(k\theta_c) + k\frac{h}{2R}F_T\cos(k\theta_c), \end{aligned} \tag{3.69}$$

leading to

$$\underline{f}_{rb} = \left(F_N + jk\frac{h}{2R}F_T\right)e^{jk\theta_c}. \tag{3.70}$$

We can again introduce $F'_T = k\frac{h}{2R}F_T$ to write:

$$\begin{aligned} \underline{f}_{ra} &= j\left(F_N + jF'_T\right)e^{jk\theta_c}, \\ \underline{f}_{rb} &= \left(F_N + jF'_T\right)e^{jk\theta_c}. \end{aligned} \tag{3.71}$$

The two reaction forces vectors, $\underline{f}_{ra}$ and $\underline{f}_{rb}$, can be written in the rotating reference frame as

$$\underline{f}_{ri} = (F_{rid} + jF_{riq}))e^{j\omega t}, \quad i \in \{a, b\}, \tag{3.72}$$

and we have

$$F_{rid} + jF_{riq} = \underline{f}_{ri}e^{-j\omega t}, \quad i \in \{a, b\}. \tag{3.73}$$

The contact force F_T derives from the torque T produced by the motor with

$$T = R \times F_T. \tag{3.74}$$

The dynamic of the rotor. The rotor is connected to a mechanical load, which is defined by its inertia J and reaction torque T. Therefore, the rotational speed of the rotor Ω follows

$$J\frac{\mathrm{d}\Omega}{\mathrm{d}t} = T - T_r, \tag{3.75}$$

leading to

$$\Omega = \frac{1}{J}\int (T - T_r)\,\mathrm{d}t. \tag{3.76}$$

Torque–speed characteristics. TWUM imposes a torque T on the mechanical load, to make it rotate, as shown in Eq. (3.76). At the same time, the stator imposes the rotational speed Ω_{id} due to the ideal rotor contact conditions and Eq. (3.64). Therefore, Ω and Ω_{id} cannot be the same state variable, because they are defined by two different equations, and $\Omega \neq \Omega_{id}$. Here we introduce the torque–speed characteristics of a TWUM. Indeed, at constant vibration velocity, $\Omega_i d$ is constant, and the torque T increases when Ω increases. We can then propose the following relationship:

$$T = f_0\,(\Omega - \Omega_{id})\,. \tag{3.77}$$

This equation leads to linear torque–speed characteristics for the TWUM, which is acceptable for most applications. A more accurate model can be used, however, at the expense of complexity.

How to take into account a non-punctual contact condition?

The *ideal contact conditions* assume that the contact is punctual, and as a consequence, the contact pressure is written using Dirac function $\delta(\theta - \theta_c)$. This allows us, at the end, to use the rotation with angle $k\theta_c$ in Eq. (3.71). But can we still use this rotation when more realistic contact conditions occur?

Practically, TWUM uses a thin layer of elastic material on the rotor to facilitate the contact conditions. If we denote $\hat{w} = \sqrt{w_a^2 + w_b^2}$ then Eq. (3.60) leads to $w = \hat{w}\cos\left(k\,(\theta - \theta_c)\right)$. The stator penetrates into the rotor, the position of which is denoted by w_R, because $\hat{w} < w_R$, as shown in the following figure:

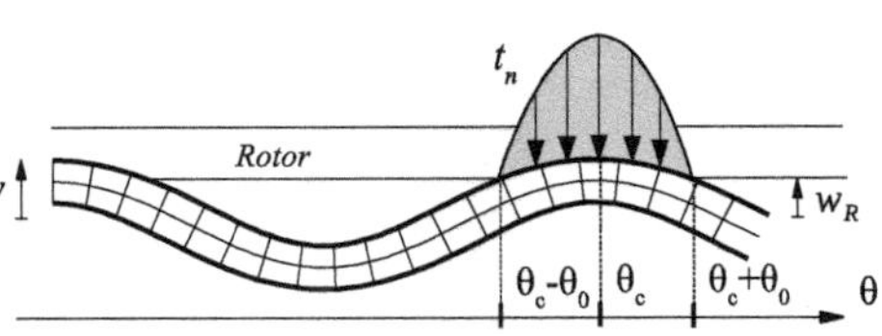

The contact is not punctual, and occurs between the angles $\theta_c + \theta_0$ and $\theta_c - \theta_0$. By definition, we have $w_R = \hat{w}\cos\left(k\,(\theta - \theta_0)\right)$. We introduce c_N, the normal stiffness of the contact layer, and write t_3, the normal force pressure, according to Eq. (3.78) as

$$t_n = c_N(w - w_R) = c_N\hat{w}\left(\cos\left(k(\theta - \theta_c)\right) - \cos(k\theta_0)\right),$$
$$-\theta_0 \le (\theta - \theta_c) \le \theta_0. \tag{3.78}$$

The value of θ_0 can be deduced from the normal force applied on the rotor, and it depends on the ratio $\frac{c_N\hat{w}}{F_N}$. Indeed, using Eq. (3.78),

$$F_N = k\int_{\theta_c-\theta_0}^{\theta_c+\theta_0} t_n d\theta = 2c_N\hat{w}\left(\sin(k\theta_0)_k\theta_0\cos(k\theta_0)\right). \tag{3.79}$$

The projection of t_n on mode a, which is called f_{na}, can be deduced by a calculation similar to that in Eq. (3.67):

$$\begin{aligned} f_{na} &= k\int_{\theta_c-\theta_0}^{\theta_c+\theta_0}\left(c_N\hat{w}\left(\cos\left(k(\theta-\theta_c)\right)-\cos(k\theta_0)\right)\right)\cos(k_theta)\mathrm{d}\theta \\ &= \frac{c_N\hat{w}}{2}\left(2k\theta_0 - \sin(2k\theta_0)\right)\cos(k\theta_c). \end{aligned} \tag{3.80}$$

Hence, compared with Eq. (3.67), f_{na} depends again on $\cos(k\theta_c)$. However, F_N doesn't show up directly in Eq. (3.80), and we introduce ε such that

$$\varepsilon = \frac{c_N\hat{w}}{2F_N}\left(2k\theta_0 - \sin(2k\theta_0)\right). \tag{3.81}$$

Therefore, the effect of the non-punctual contact can simply be taken into account by replacing F_N by εF_N in Eq. (3.71). Moreover, ε depends on the ratio $\frac{c_N\hat{w}}{F_N}$, and we show its evolution in the following figure:

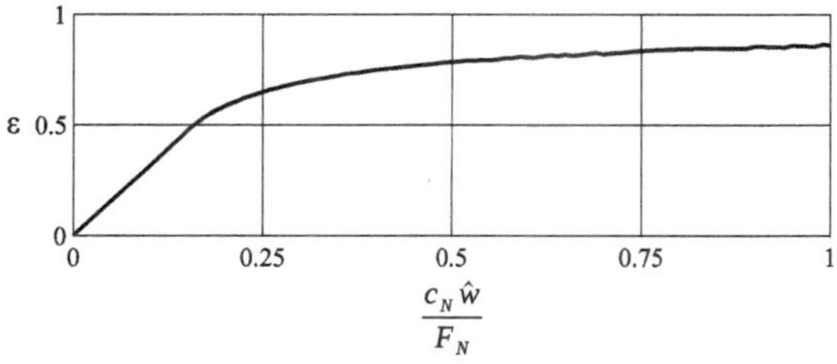

The same calculation can be made for phase b, and we obtain the same conclusion: F_N should be replaced by εF_N to take into account the non-punctual contact condition. Of course, this non-ideal contact condition also occurs on the tangential axis; but it is much harder to calculate its effect on the projection of F_T, and a ratio like ε is not so obvious to introduce.

3.4.4.2 EMR of the TWUM with two independent phases

All the aforementioned equations can be represented using the EMR blocks of Tables 3.1 and 3.2. The result is given in Fig. 3.24, which should replace the 4 mechanical sources of Fig. 3.16.

Theoretically, it is possible to choose θ_1 and θ_2 in such a way that the exciter E_1 controls mode a and exciter E_2 controls mode b. This is achieved with $\theta_1 = 0 + n\frac{\pi}{k}$ and $\theta_2 = \frac{\pi}{2} + n\frac{\pi}{k}$, $n \in \mathbb{N}$. The matrix Φ_{ab} takes then a particular form:

$$\Phi_{ab} = \begin{pmatrix} 1 & 0 \\ 0 & 1 \end{pmatrix}. \tag{3.82}$$

Therefore, the couplings between phases 1 and 2 do not exist in this simulation. We then simulate the motor at constant voltage amplitude, and vary pulsation ω with

$$v_1 = -V\cos(\omega t) \quad \text{and} \quad v_2 = V\sin(\omega t). \tag{3.83}$$

Simulations are carried out with the parameters from Table 3.4, which correspond approximately to those of a Shinsei USR60 presented in Chap. 1. In this part, we consider that the motor is unloaded, *i.e.*, $T_r = 0$.

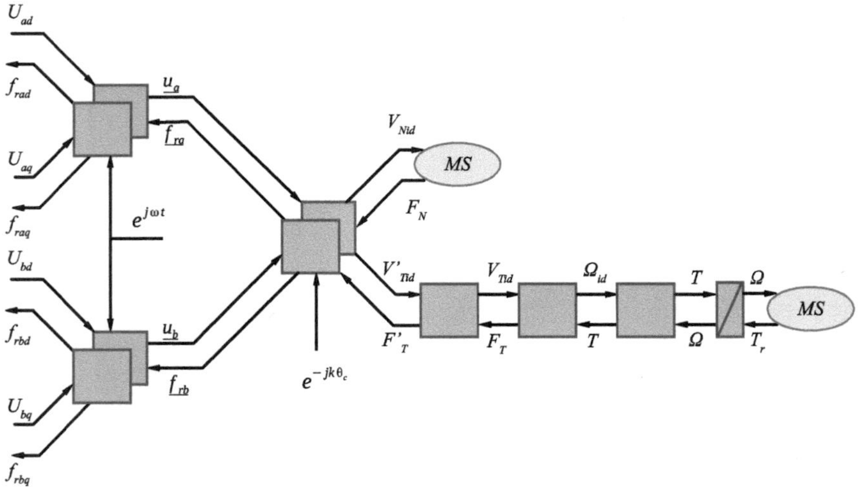

Figure 3.24 EMR of the mechanical part of a TWUM; the blocks replace the 4 mechanical sources of Fig. 3.11.

Table 3.4 Parameters values of a Shinsei USR60, taken from (Li et al., 2018).

M	D	K	N	h	R
22 g	32 Nsm^{-1}	1.4210^9 Nm^{-1}	0.32 N/V	3 mm	30 mm
F_N	k	c_N	f_0	J	V
260 N	9	0.756×10^9 Nm^{-1}	0.67 Nms	10^{-4} $\mathrm{kg\,m}^2$	200 V

This gives rise to the frequency behavior of the stator, as depicted in Fig. 3.25 for 3 levels of normal force: $F_N = 260$ N, which corresponds to the nominal loading, $F_N = 130$ N and $F_N = 0$ N.

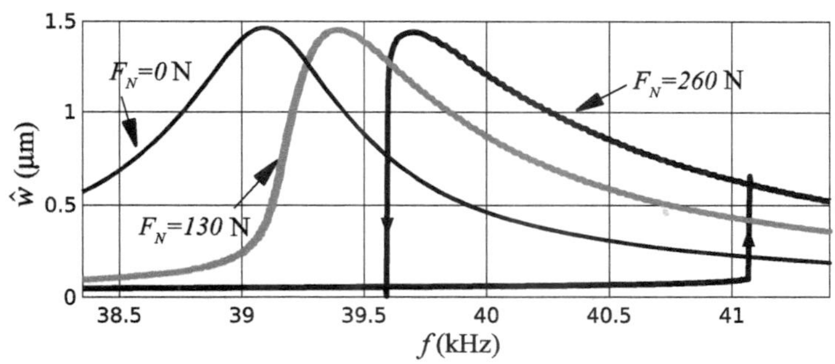

Figure 3.25 Vibration amplitude $\hat{w}$ as a function of $f = \omega/2\pi$ for several levels of normal loading.

The results clearly show a *pull-out* phenomenon, more visible at $F_N = 260$ N than at $F_N = 130$ N, while it doesn't exist at $F_N = 0$ N. Moreover, it is almost impossible to start-up the motor with increasing frequency at the nominal normal force, and operation should be carried out by decreasing the frequency. This behavior is typical of the TWUM, and it is well documented in the literature (Li et al., 2018). Hence, the *pull-out* phenomenon in TWUM is not due to nonlinearity of the material used, but rather it is due to the high normal loading.

Fig. 3.26 shows simulations of the motor at different values of the voltage amplitude V, keeping $F_N = 260$ N.

These results show that due to the pull-out phenomenon, the resonant frequency increases when the voltage decreases. Therefore, operation at variable voltage amplitude of TWUM, as it will be presented in Chap. 5, should include the tracking of the resonance frequency to operate at the best conditions. However, below 100 V, the simulations show that the vibration amplitude $\hat{w}$ is very limited: the motor cannot start-up. Hence, it is necessary to supply a minimal voltage to the motor.

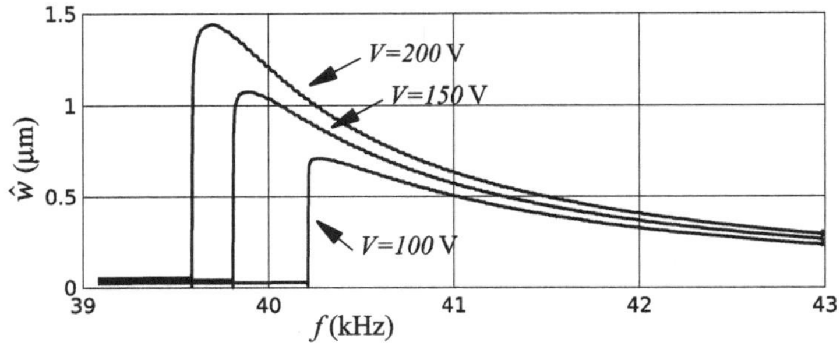

Figure 3.26 Vibration amplitude $\hat{w}$ as a function of $f = \omega/2\pi$ for several levels of voltage amplitude V.

More surprisingly, once the motor has started, it is not necessary to maintain a voltage on the motor to have a rotation. For instance, Fig. 3.27 shows what happens when we suddenly set $v_1 = 0$ at $t = 0$:

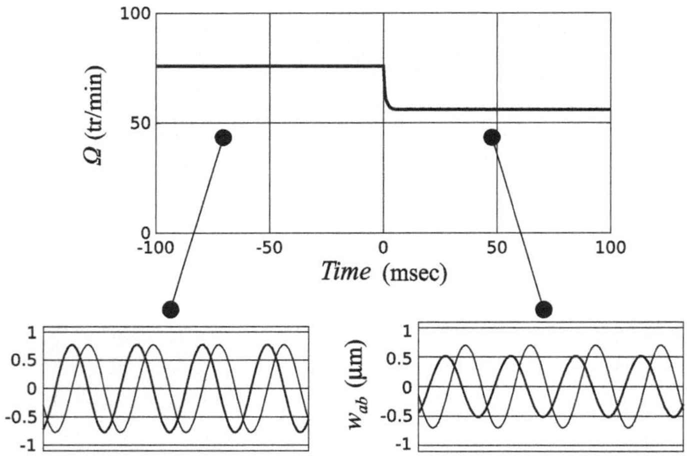

Figure 3.27 Evolution of the motor's speed Ω as a function of time and the corresponding vibration displacements w_a and w_b; at $t = 0$, v_1 is suddenly switched to 0.

Hence, once the motor has started-up, removing one voltage doesn't stop it, and the motor continues its rotation. The evolution of w_a and w_b show that $w_a \neq 0$ even if $v_1 = 0$. This is possible because the modes a and b are coupled with the rotor by the rotation $e^{jk\theta_c}$, as illustrated by the representation in Fig. 3.22 and some authors use this coupling to simplify the power supply (Petit and Gonnard, 2004). However, this new operating condition changes the vibration of the stator, and the speed V'_{Tid} is reduced with one phase only: the motor has decelerated, but hasn't stalled.

If the motor is loaded, with the nominal operating conditions, then the motor's speed is decreased, as depicted in Fig. 3.28, where a load torque of 0.25 Nm is applied at $t = 0$.

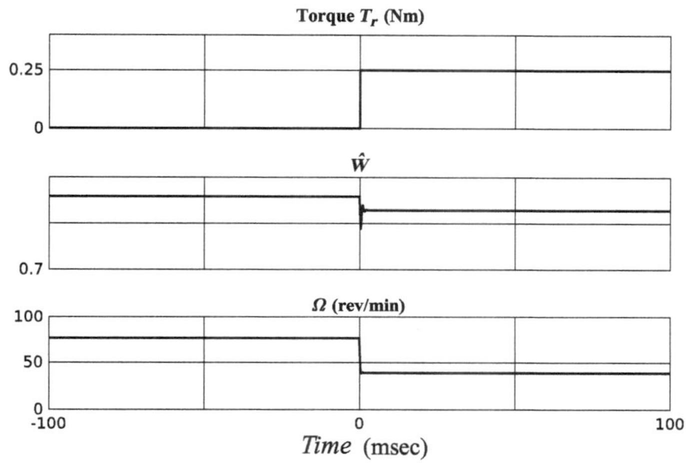

Figure 3.28 Evolution of the motor's speed Ω as a function of time, and the vibration amplitude $\hat{W}$ when a load torque is applied on the motor's shaft.

There are 2 reasons why the motor's speed decreases:

- First, due to the torque/speed characteristics and coefficient f_0, applying torque leads to a difference between Ω_{id} and Ω;
- Second, the load torque has an influence on the vibration, and acts like an additional damping in some circumstances; thus, the vibration amplitude decreases, which makes Ω_{id} decrease.

This model approximates the behavior of the TWUM; it can take into account some features like the effect of the normal force, or the torque–speed characteristics. In the next section, we simplify this model to learn some aspects of its behavior better, leading to a definition of its operating area.

3.4.4.3 Simplified model

To simplify the model, we consider that the motor has no coupling between phases 1 and 2, and the two vibration modes have the same parameters. Moreover, we assume that they vibrate with the same amplitude, with a phase shift of $\frac{\pi}{2}$:

$$w_a(t) = W\cos(\omega t - \varphi) \quad \text{and} \quad w_b(t) = W\sin(\omega t - \varphi). \qquad (3.84)$$

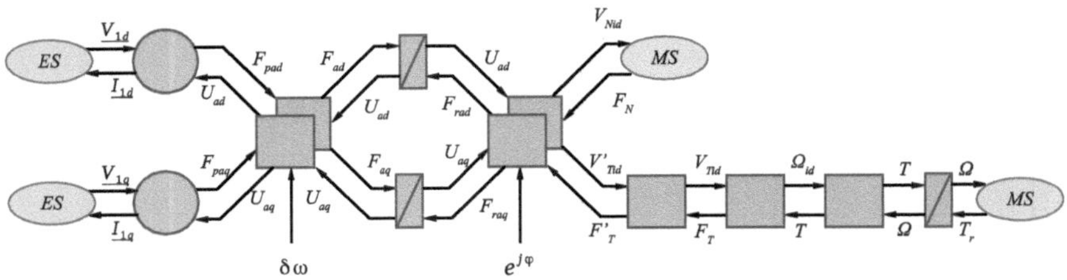

Figure 3.29 Equivalent 1-phase, 1-mode model of a TWUM in the rotating reference frame.

This assumption is typical, and Eq. (3.84) actually describes the more efficient way to operate a TWUM. Due to Eq. (3.19), we remark that Eq. (3.84) yields:

$$\begin{aligned} U_{da} &= -U_{qb} = -\omega W \cos(\varphi), \\ U_{qa} &= U_{db} = \omega W \sin(\varphi). \end{aligned} \tag{3.85}$$

What is interesting is that in this way, the rotating reference frame is "attached" to the contact point, because Eq. (3.62) gives

$$k\theta_c = \omega t - \varphi. \tag{3.86}$$

As a consequence, the two rotations of Eqs. (3.71) and (3.73) compensate to eventually give a constant rotation of $-\varphi$; these equations simplify into:

$$\begin{aligned} F_{rad} + jF_{raq} &= j\left(F_N + jF'_T\right) e^{j\varphi}, \\ F_{rbd} + jF_{rbq} &= \left(F_N + jF'_T\right) e^{j\varphi}. \end{aligned} \tag{3.87}$$

We can then replace the model of the TWUM by using one single phase, like phase 1, which also directly controls mode a, by the model of a transducer in the rotating reference frame, where we have Eq. (3.87) to calculate the reaction forces. This gives rise to the model of Fig. 3.29.

This model will be used in Chap. 5 to control a TWUM efficiently. But to analyze its behavior, we fix the reference frame on the vibration, by imposing $U_{qa} = 0$ and $\varphi = \pi$. Then, Eq. (3.87) leads to $F_{rad} = F'_T$ and $F_{raq} = -F_N$.

Introducing this result into Eq. (3.41), and for the steady state, we obtain:

$$\begin{aligned} DU_{ad} &= F_{pd} - F'_T \rightarrow F_{pd} = DU_{ad} + F'_T, \\ 2M\delta\omega U_{ad} &= F_{pq} + F_N \rightarrow F_{pq} = 2M\delta\omega U_{ad} - F_N. \end{aligned} \tag{3.88}$$

Therefore, from Eq. (3.88), we can conclude that force f_p and thus voltage v_1 is split into 2 parts:

- along axis d, it compensates the torque produced by the motor F'_T and the vibration losses DU_d,
- along axis q, it compensates the normal force F_N and the coupling $2M\delta\omega U_d$.

This is why, at constant pulsation ω and voltage amplitude, increasing the torque on the motor decreases the vibration velocity: the voltage on axis d is shared in one part for the vibration, the other is used for the torque. If the torque is increased, less voltage is available for the vibration velocity. This is explained by Fig. 3.30.

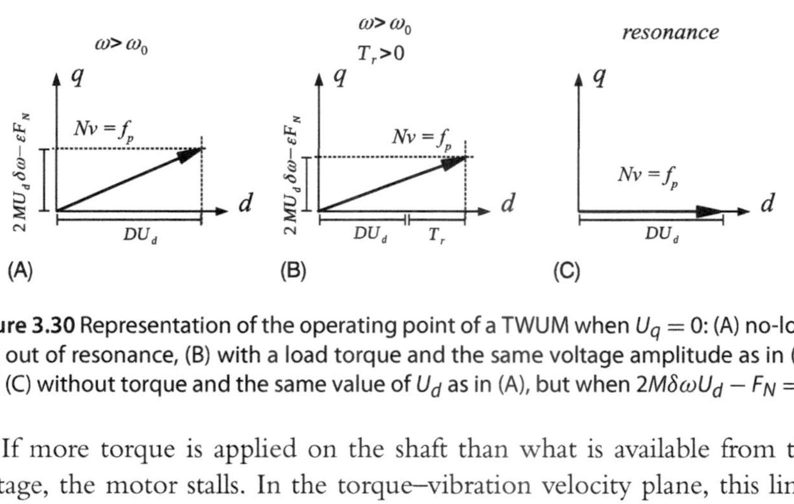

Figure 3.30 Representation of the operating point of a TWUM when $U_q = 0$: (A) no-load and out of resonance, (B) with a load torque and the same voltage amplitude as in (A), and (C) without torque and the same value of U_d as in (A), but when $2M\delta\omega U_d - F_N = 0$.

If more torque is applied on the shaft than what is available from the voltage, the motor stalls. In the torque–vibration velocity plane, this limit is obtained from Eq. (3.88); the motor can operate if

$$U_d < \frac{NV - k\frac{h}{2R^2}T_r}{D}. \tag{3.89}$$

This limit is summarized in Fig. 3.31.

Finally, this simplified model is helpful to capture why the resonance frequency shifts due to the normal force. Indeed, the resonance occurs when the voltage amplitude is the smallest for a given vibration velocity.

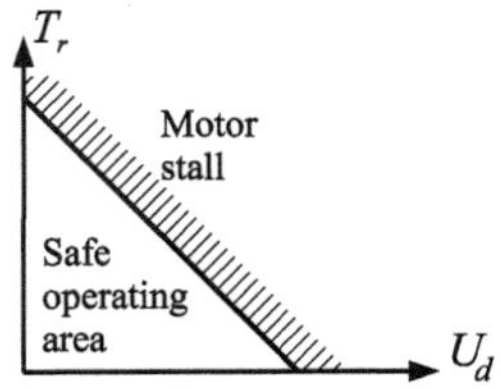

Figure 3.31 Safe operating area of a TWUM.

This happens when $2MU_d\delta\omega - \varepsilon F_N = 0$. Hence, $\omega = \omega_0$ is not the condition of the resonance, and, moreover, the operating pulsation should be adapted to the vibration velocity to operate at resonance. In Chap. 5, we will use this conclusion to achieve efficient control of a TWUM in the rotating reference frame.

CHAPTER FOUR

Control in the rotating reference frame

Contents

4.1. Introduction 138
4.2. Control schemes 138
4.2.1 Methodology 138
4.2.2 Control of the vibration speed of an ultrasonic transducer 140
4.2.2.1 Control of the vibration velocity at constant frequency 140
4.2.2.2 Tracking the resonance frequency 141
4.2.3 Control of a traveling wave 144
4.3. Controller design 145
4.3.1 Simulation of a Langevin transducer controlled in vibration, without frequency tracking 146
4.3.1.1 Robustness with respect to load change 148
4.3.1.2 Robustness with respect to resonance pulsation estimation 148
4.3.2 Simulation of the frequency tracking of a Langevin transducer 152
4.3.2.1 Robustness of the frequency tracking with respect to variation of the load damping 155
4.3.2.2 Robustness of the frequency tracking with respect to variation of the load equivalent stiffness 155
4.4. Observers 157

Abstract

The objective of this chapter is to present the control schemes which derive from the modeling in the rotating reference frame. We show how to control the amplitude of the vibration speed of a transducer, track its resonance frequency, and synchronize two transducers in order to perfectly control a traveling wave in a finite beam. The control schemes are obtained after following a methodology that consists in inverting the previously obtained model.

We identify the need of vibration speed controllers, and the synthesis of these is given afterwards. Simulation results are presented in order to clearly show how these controls operate. Moreover, in the last section, we implement observers in order to estimate the acoustic impedance of the load.

All the control schemes are given in the rotating reference frame, thus achieving the vector control of the transducers.

Piezoelectric Actuators: Vector Control Method
https://doi.org/10.1016/B978-0-12-814186-1.00008-9

Keywords

control in rotating reference frame, ultrasonic transducer, vibration velocity, resonance frequency, controller design, robustness, observer

4.1. Introduction

The closed-loop control of a system offers the possibility to dynamically adjust its inputs in order to follow a reference set point, and to reduce the effect of the perturbations from the system's environment. Indeed, we have seen that nonlinear behavior can occur, like the *pull-out* phenomenon for instance. Closed-loop control is achieved by adding more complexity to the system, and this makes it more vulnerable: Digital controllers have to be added and tuned, so are sensors, which are needed to close the loop. But the benefits in terms of robustness and improved performance, e.g., efficiency, always justify the cost, provided that a clear and systematic methodology can be adopted to design the control.

The section dedicated to the model has shown how similar the piezoelectric transducers are from this point of view, while they can be used for totally different applications. We will see that the *vector control method* (VCM), which consists in controlling the vibration in the rotating reference frame, can indeed produce versatile and robust control schemes, which can be adopted in many applications.

In this chapter, our objective is to control the vibration speed u. Moreover, we will consider that the voltages V_d and V_q are perfectly mastered, and that we are able to extract U_d and U_q from the measurement of u. The practical details on how to achieve this will be given in Chap. 5.

4.2. Control schemes

4.2.1 Methodology

The methodology followed consists in creating a control scheme by inverting the model of the system. For this we apply the design rules of the Energetic Macroscopic Representation. Indeed, the EMR depicted in Fig. 3.10 on page 112 emphasizes the causality in the system, and we can highlight the action chain, *i.e.*, the chain between the system's inputs (here the voltage and $\delta\omega$) to the system's output (here the vibration speed). By following step by step this action chain, we reach a control scheme by inverting each block of this chain.

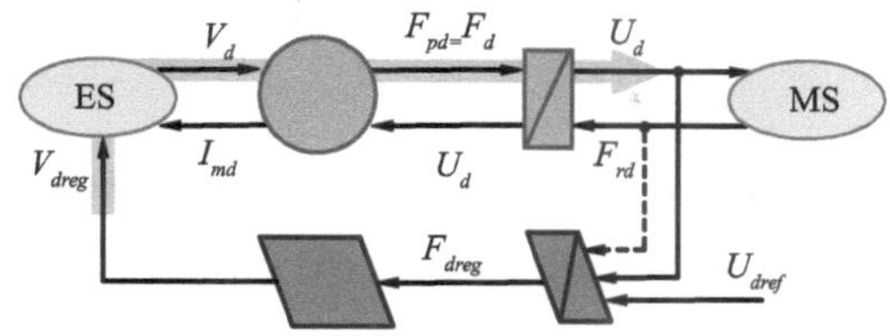

Figure 4.1 Control scheme of a simplified model in axis *d*.

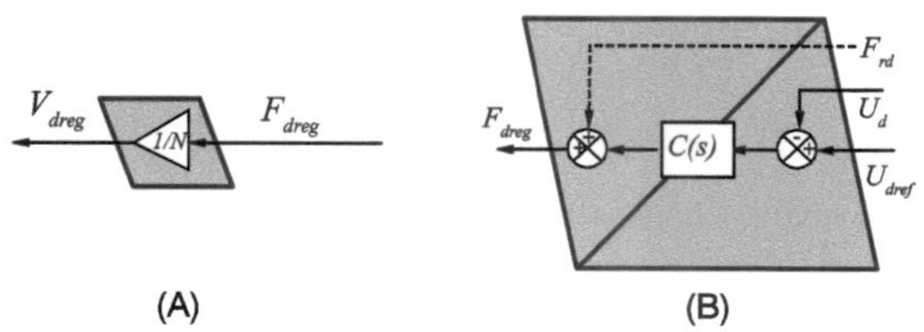

Figure 4.2 Inversion of a conversion element (A) and an energy storage element (B).

For the purpose of illustration, we apply this methodology to a simplified model presented in Fig. 4.1. It is based on the model in axis d of a Langevin transducer, operating at resonance such that $\delta\omega = 0$. So we have removed the coupling between axes d and q, leading to $F_d = F_{pd}$. The model is represented in orange (mid gray in print version), the action chain in yellow (light gray in print version), and the control scheme is in blue (dark gray in print version). We suppose that the voltage V_d is controlled by the electrical source directly in the rotating reference frame; therefore, an input is added to the electrical source labeled V_{dreg} and we have $V_d = V_{dreg}$.

The action chain starts from V_{dreg}, and to reach U_d on the action path, it goes through the first conversion stage (the piezoelectric energy conversion), and a storage element.

The control scheme consists in inverting the action path, first by drawing the inverted elements in blue (dark gray in print version), as depicted in Fig. 4.1. The conversion element can be inverted directly: because $F_{pd} = NV_d$, we can calculate $V_{dreg} = \frac{1}{N}F_{dreg}$, leading to the representation of Fig. 4.2.

The energy storage cannot be directly inverted. Indeed, since the model gives $U_d = \frac{1}{2M}\int (F_d - F_{rd} - DU_d)\,\mathrm{d}t$, the direct inversion yields $F_d = F_{rd} + DU_d + 2M\dot{U}_d$. This inversion is not causal, because it uses the derivative of U_d. Practically, differentiating U_d is difficult to achieve, and prone to noise in the measurement of U_d. Therefore, we use a closed-loop control of U_d

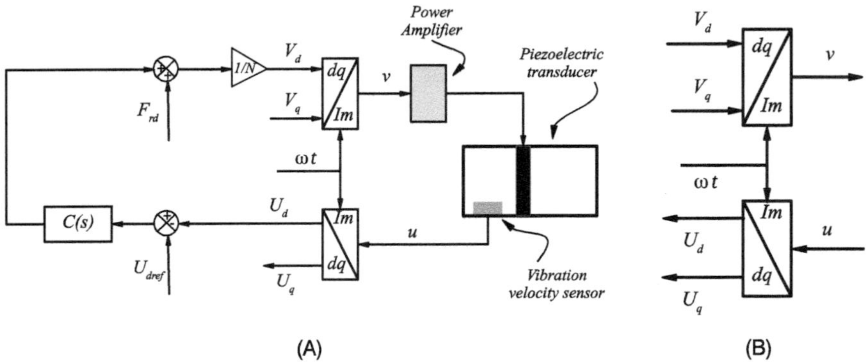

Figure 4.3 Control scheme which derives from the model inversion in axis *d* (A) and symbols of the two transformations used (B); **Im** means imaginary part in this context.

instead, and we will write:

$$F_{dreg} = F_{rd} + C(s)\left(U_{dref} - U_d\right). \tag{4.1}$$

This inversion is given in Fig. 4.2 (B), and we detail in Fig. 4.3 the practical control scheme of U_d. In this figure, we used the symbols of Fig. 4.3 (B) to represent the two transformations which transform V_d and V_q into the voltage v and measurement of u into U_d and U_q. The practical implementation of these transformations will be detailed in Chap. 6.

This methodology will be followed in the next section; the controller C will be designed in Sect. 4.4.

4.2.2 Control of the vibration speed of an ultrasonic transducer

To control the vibration of the transducer, we have to control U_d and U_q, so we have two objectives. To achieve them, we have three inputs: V_d, V_q and $\delta\omega$, which is one too many. How we choose these two variables to control the transducer defines a strategy. There are several, and we detail three of them below.

4.2.2.1 Control of the vibration velocity at constant frequency

In some circumstances, the operating frequency is fixed by the application. Therefore, $\delta\omega$ is a constant which cannot be changed. The inversion of the representation of the system highlights two control chains, one from

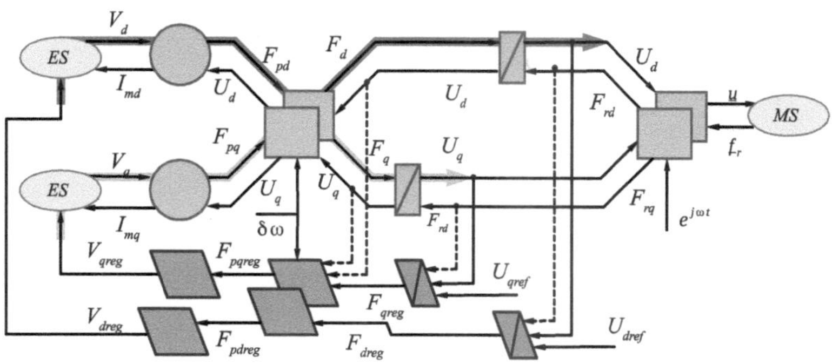

Figure 4.4 Control of U_d and U_q when the frequency is fixed.

V_{dreg} to U_d, the other from V_{qreg} to U_q, as depicted in Fig. 4.4. The control scheme is obtained by the inversion of the control chains.

Fig. 4.4 shows that to control the vibration amplitude, we need two vibration controllers, one for each axis. Their outputs are the forces F_{dreg} and F_{qreg} which cannot be set directly. Indeed, they derive from the coupling between the axes d and q, and this coupling has to be inverted, too. This is why we have introduced the block depicted in Fig. 4.5 (A). Since

$$F_d = F_{pd} + 2MU_q\delta\omega, \tag{4.2}$$

the coupling inversion will lead to

$$F_{pdreg} = F_{dreg} - 2M\delta\omega U_q. \tag{4.3}$$

The same methodology is applied in axis q, leading to

$$F_{pqreg} = F_{qreg} + 2M\delta\omega U_d. \tag{4.4}$$

We then arrive at the compensation of the coupling; and its Simulink implementation is depicted in Fig. 4.5 (B).

Finally, Fig. 4.6 shows the control scheme of the transducer in the rotating reference frame, which derives from the inversion of the model. This diagram takes up a large number of the elements of the inverted EMR. How we implement this scheme in a real time controller will be studied in Chap. 6.

4.2.2.2 Tracking the resonance frequency

As it has been described in Chap. 1, the resonance frequency of a piezoelectric transducer shifts due to nonlinearities and/or temperature variation.

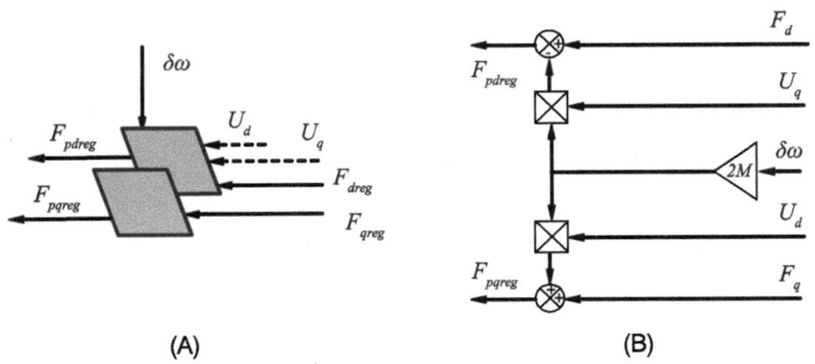

Figure 4.5 Simulink implementation of the coupling compensation when the frequency is fixed.

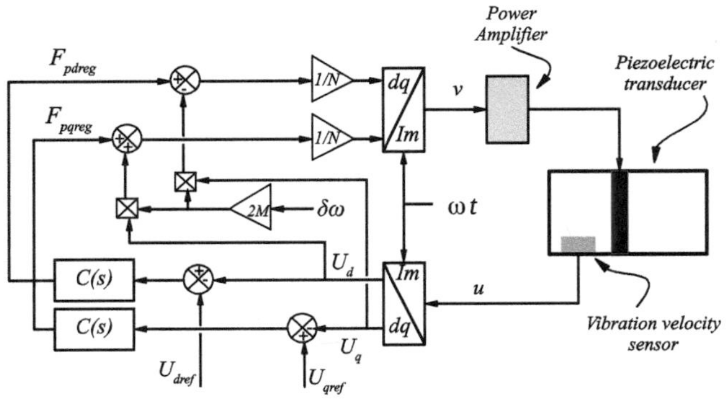

Figure 4.6 Control scheme of the vibration in the rotating reference frame when the frequency is fixed.

Therefore, it is essential to operate at resonance in order to preserve the efficiency of the system. But how can we track the resonance frequency?

We noticed in Sect. 3.3.2 that at resonance, we have $U_q = 0$ when $V_q = 0$. Starting from a predefined operating pulsation that we call ω_i, the frequency tracking algorithm then consists in adjusting the pulsation until $U_q = 0$ while maintaining $V_q = 0$.

In our modeling, the system's input, which corresponds to the pulsation, is $\delta\omega$. This is why we depict in Fig. 4.7 the action chain from $\delta\omega$ to U_q.

The control scheme consists then of two vibration speed controllers, one in axis d, the other in axis q. These controllers can be the same as those designed for the case operating at constant pulsation. The outputs of these controllers are F_{dreg} and F_{qreg}. The inversion of the cross-coupling is achieved by inverting the coupling rules.

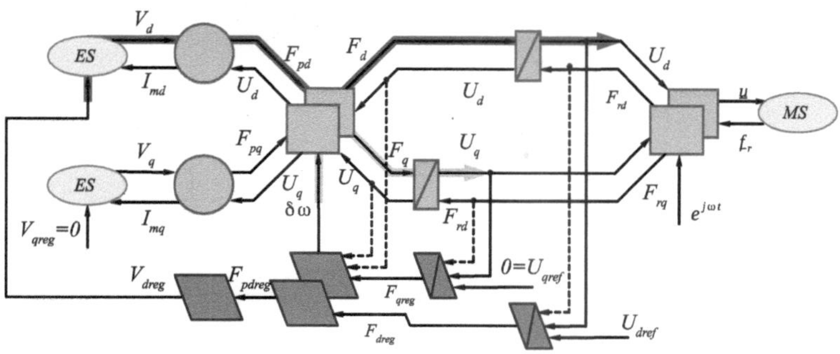

Figure 4.7 Control scheme of a transducer with automatic frequency tracking.

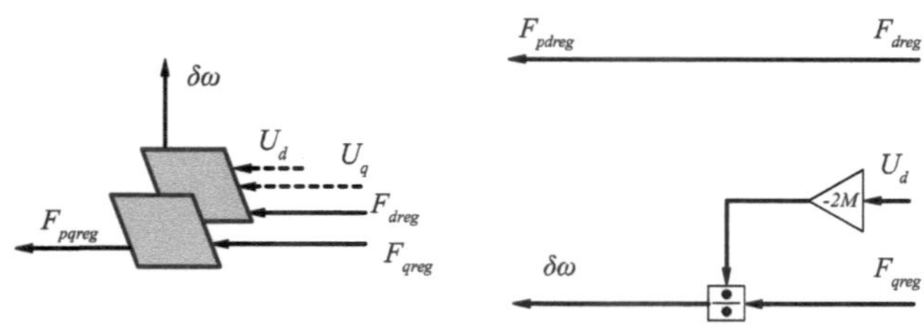

(A) (B)

Figure 4.8 Simulink implementation of the couplings compensation for the frequency tracking.

The controller in axis q calculates the reference for F_q denoted F_{qref}. But F_q cannot be controlled with V_q because $V_q = 0$. Therefore, we use the coupling and $\delta\omega$ to control F_q. Indeed, we have

$$F_q = F_{pq} - 2MU_d\delta\omega = -2MU_d\delta\omega, \tag{4.5}$$

because $V_q = 0$. This is why the inversion of the coupling gives for the axis q the following expression:

$$\delta\omega = -2M\frac{F_{qreg}}{U_d}. \tag{4.6}$$

For d axis, since U_q is controlled to 0, there is no coupling, and (4.3) leads to $F_{pdreg} = F_{dreg}$. Finally, a Simulink implementation of the compensation of the couplings is depicted in Fig. 4.8.

The scheme of this control using blocks is given in Fig. 4.9.

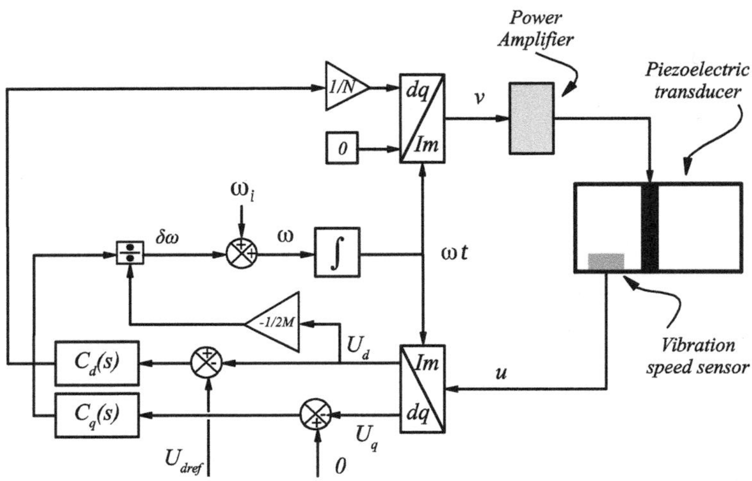

Figure 4.9 Control scheme of the vibration in the rotating reference frame with frequency tracking.

4.2.3 Control of a traveling wave

To obtain a traveling wave, we have shown in Chapter 3 that the vibration of the two chosen successive modes needs to have the same amplitude shifted by 90°. In the rorating reference frame, this leads to the control of four state variables (U_{ad}, U_{aq}, U_{bd}, U_{bq}). For these objectives, we have five inputs (V_{1d}, V_{1q}, V_{2d}, V_{2q} and ω). In this application, following a resonance frequency is useless, because it is necessary to operate between the resonance pulsation of the two modes. Therefore, the control of the traveling wave is achieved at fixed pulsation as achieved by the control scheme depicted in Fig. 4.4, and which guarantee that the vibration velocity are controlled even if the environment condition (like temperature for instance), make the resonance pulsation of one mode get closer to the working pulsation. However, since the two exciters of a vibrating beam (or a stator of a Traveling Wave Ultrasonic Motor) contribute to the vibration of each mode, the controller needs to share of the force acting on each mode on each actuator. This is achieved by the control depicted in Fig. 4.10, and which is obtained by inverting the control chains in the EMR depicted in Fig. 3.16 on page 120.

The control scheme shows that the control of the vibration speeds $\underline{U}_a$ and $\underline{U}_b$ is similar to that of an ultrasonic transducer, and produces the references for $\underline{F}_{pa}$ and $\underline{F}_{pb}$. Using Φ_{12}^{-1}, the inverse of the coupling matrice Φ_{12},

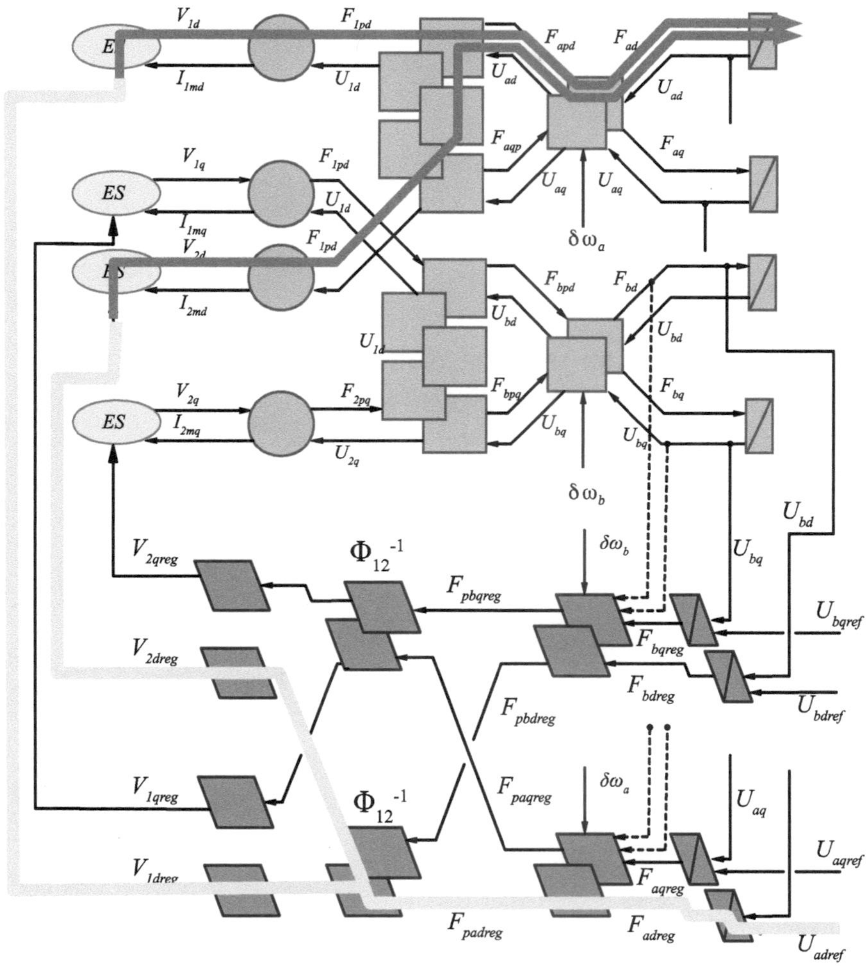

Figure 4.10 EMR of the control of a traveling wave in a rotating reference frame; tuning paths are presented for the control of $\underline{U}_{ad}$.

these references are then decoupled in order to give the force reference for each actuator, denoted by $\underline{F}_1$ and $\underline{F}_2$, leading to $\underline{V}_1$ and $\underline{V}_2$.

This control will be experimentally checked in Chapter 5.

4.3. Controller design

In this section, we design the controllers and test the robustness of the closed-loop control by simulations. We apply the aforementioned strategies

Table 4.1 Parameters of a Langevin transducer used for the simulations.

M	D	K	N
0.249 kg	58.4 Ns/m	$7.64 \cdot 10^9$ N/m	0.45 N/V

in three different cases: the Langevin transducer controlled in vibration speed and amplitude; and two piezoelectric actuators. The parameters used for the simulation are given in Table 4.1.

4.3.1 Simulation of a Langevin transducer controlled in vibration, without frequency tracking

In order to design the two controllers for axes d and q, we start from the Open Loop Transfer Function (OLTF, see Sect. 1.4, page 32). For this we assume that the couplings are compensating, leading to $F_d = F_{dreg}$.

From the energetic macroscopic representation of the model, limited to the energy storage block to be inverted, we can draw in Fig. 4.11 the block diagram of the closed-loop control of the vibration speed amplitude, and in such a way that it is similar to the diagram already studied in Sect. 1.4.

If we can precisely compensate for the perturbation of the load, by measuring F_{rd} or by estimating it, or if it is null because the actuator is not

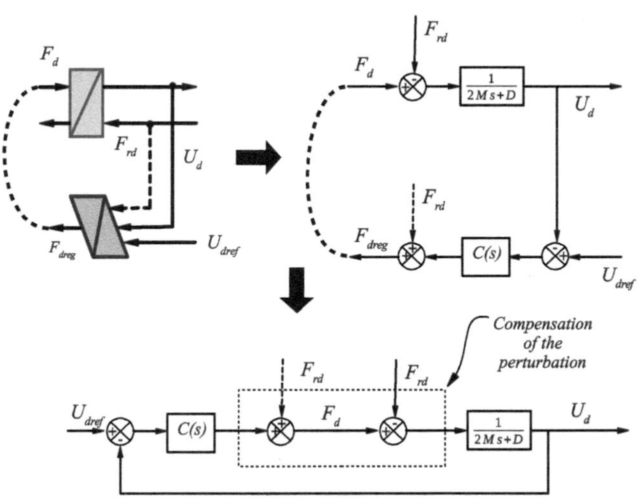

Figure 4.11 Block diagram of the closed-loop control of the vibration speed amplitude in axis *d*.

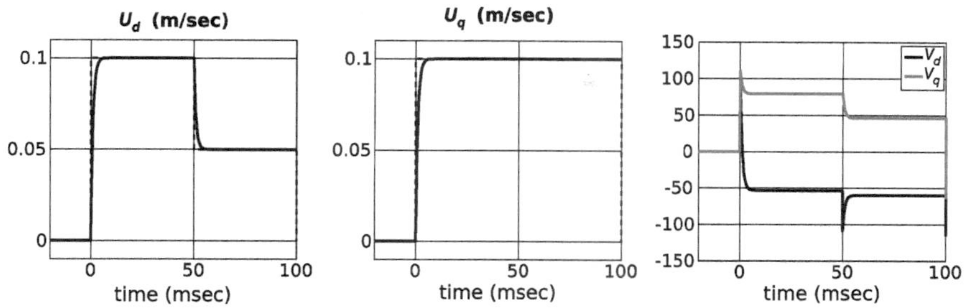

Figure 4.12 Transient response of U_d and U_q in the rotating reference frame, and the corresponding voltages V_d and V_q.

loaded, the *OLTF* is equal to

$$OLTF = C(s)\frac{1}{2Ms + D}. \tag{4.7}$$

We notice that the system shows only one pole at $\tau = \frac{2M}{D}$ while the gain equals $\frac{1}{D}$. Therefore, we can use a PI controller in order to cancel the static error, and will have

$$C(s) = K_p + \frac{K_I}{s}. \tag{4.8}$$

The controller is designed to compensate the dominant pole, and we chose a closed-loop response time of $t_r = 3$ ms. In theory, t_r can be set to any value, and we usually set the lowest values which are compatible with the hardware. However, t_r is limited by the *dynamic assumption* – which in our case leads to $t_r \gg \frac{1}{\sqrt{\frac{K}{M}}} = 5$ µs. These settings lead to (see page 37):

$$K_p = \frac{3 \times D \times \frac{2M}{D}}{t_r} = 498 \quad \text{and} \quad K_i = \frac{2M}{D} \times K_d = 4.24. \tag{4.9}$$

Since the model in q axis is equal to that in d axis, this controller can be set in q axis as well.

We present simulation results in Fig. 4.12. Starting from 0, the reference set-points are 0.1 m/s for U_d and U_q, then at $t = 50$ ms, U_{dref} is set to 0.05 m/s. In this simulation, we consider $\delta\omega = 600$ rad/s, in order to illustrate the effects of the couplings between d and q.

The vibration speeds U_d and U_q follow their references, and the response time is respected; the voltages V_d and V_q show a spike at $t = 0$ to overcome the transducer's natural dynamic, and then stabilize at steady state. The fact that the values of V_d and V_q are different, while U_d and U_q are equal, is due to the couplings which operate differently in axes d and q. At $t = 50$ ms, the change in U_d does not affect U_q, and this is a result of the decoupling shown in Fig. 4.6. The voltages V_d and V_q automatically adapt, and we can notice that V_d changes less than V_q: this is because the decoupling the two axes requires more voltage than creating the vibration for the pulsation shift $\delta\omega$ considered here.

4.3.1.1 Robustness with respect to load change

The acoustic load which is connected to the system is modeled by f_r. For example, we consider that the load behaves like a damping, with $f_r = \tilde{D}u$, leading to

$$F_{rd} = \tilde{D}U_d \quad \text{and} \quad F_{rq} = \tilde{D}U_q. \tag{4.10}$$

In the simulation results presented in Fig. 4.13, the load is activated at $t = 50$ ms, with a variation equal to 100% of D ($\tilde{D} = 0$ if $t < 50$ ms, and $\tilde{D} = D$ if $t > 50$ ms), while $U_{dref} = U_{qref} = 0.1$ m/s. In the first try, the compensations of F_{rd} and F_{rq} are set (see Fig. 4.11). In the second try, this compensation is removed.

When the perturbation of the load is compensated, it has no consequence on the vibration speed of the transducer, because the voltages are automatically increased. This appears in the ellipse of Fig. 4.13. But when this compensation is removed, the *PI* controller has to compensate for the perturbation. When the perturbation appears, U_d and U_q suddenly drop and the errors $U_{iref} - U_i$ increase ($i \in \{d, q\}$). The integrators' outputs of the *PI* controllers increase, so do the voltages V_d and V_q, until the perturbation is compensated for. This is highlighted in Fig. 4.13 (B) with ellipses.

It is difficult to measure the force f_r on the device, and most of the time, the compensation is not achieved. Therefore, sudden changes of the load will create variations of the vibration speed amplitude. However, these variations are limited: in our example, the vibration amplitude drops by less than 10%, and the system recovers after 25 ms. If more performances are required, a solution can be having a load estimator to perform the compensation.

4.3.1.2 Robustness with respect to resonance pulsation estimation

When decoupling the two axes, it is necessary to know the value of $\delta\omega$, the difference between the operating and resonance pulsations. However,

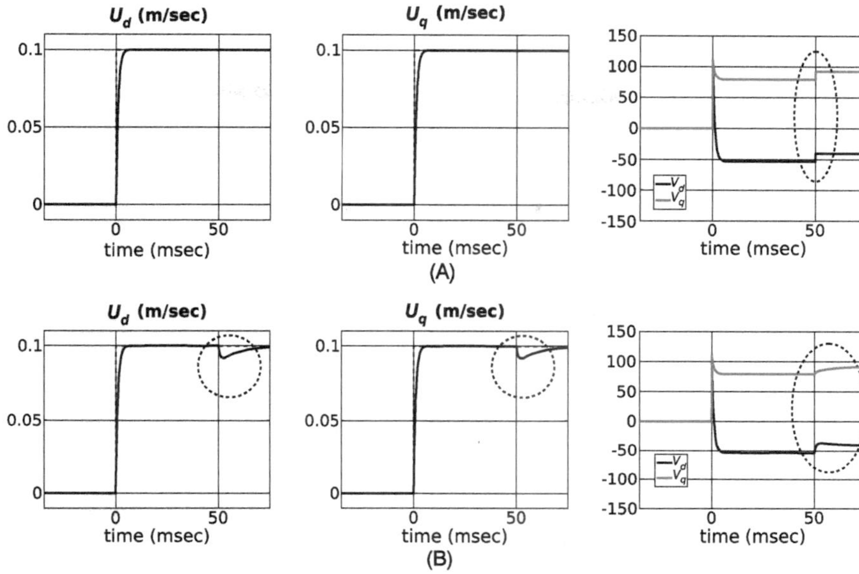

Figure 4.13 Effect of the compensation of f_r on the robustness: (A) with compensation for the perturbation, (B) without compensation.

as it has been highlighted in previous chapters, the resonance pulsation is not perfectly known, and varies with the system's temperature or with the nonlinearities of the materials. Therefore, the coupling compensations are often imperfect. To illustrate the effect of an incorrect decoupling of the two axes, we depict in Fig. 4.14 the system's response to a step reference, when $\delta\omega = 600$ rad/s, but having it estimated as $\delta\omega = 400$ rad/s for the compensation.

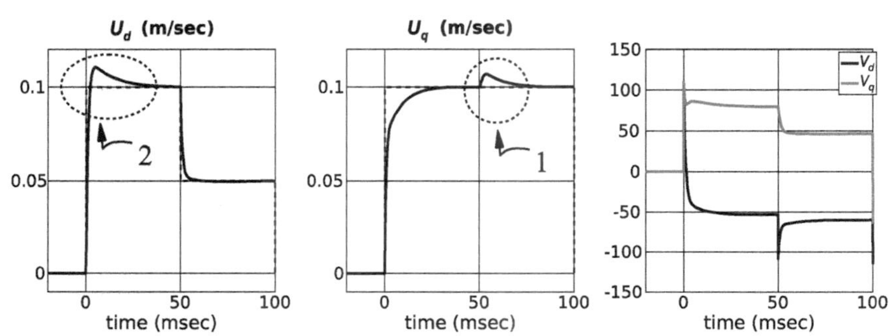

Figure 4.14 Effect of an estimation error of $\delta\omega$ on the transient response of the vibration.

Consequences of an imperfect decoupling on $\frac{U_d}{F_{dreg}}$

We denote by $\widehat{\delta\omega}$ the estimated value of $\delta\omega$ such that $\widehat{\delta\omega} = \delta\omega + \Delta$, and want to calculate the transfer function $\frac{U_d}{F_{dreg}}$ when $\Delta \neq 0$. Eqs. (4.3) and (4.5) yield

$$F_{pdreg} = F_{dreg} - 2M\widehat{\delta\omega}U_q \quad \text{and} \quad F_{pqreg} = F_{qreg} + 2M\widehat{\delta\omega}U_d \tag{4.11}$$

If we consider that there is no error in the conversion inversion, we have $F_{pd} = F_{pdreg}$, and that, for the sake of simplicity and without loss of generality, $F_{rd} = F_{rq} = 0$, the model in axis d and Eq. (3.43) on page 112 give

$$\begin{aligned} U_d &= \frac{1}{2Ms + D}\left(F_{dreg} - 2M\Delta U_q\right) \quad \text{and} \\ U_q &= \frac{1}{2Ms + D}\left(F_{qreg} + 2M\Delta U_d\right). \end{aligned} \tag{4.12}$$

Eliminating U_q in the first equation leads to

$$U_d\left(1 + \frac{1}{(2Ms + D)^2}\left(4M^2\Delta^2\right)\right) = \frac{1}{2Ms + D}\left(F_{dreg} - \frac{2M\Delta}{2Ms + D}F_{qreg}\right). \tag{4.13}$$

The transfer function between F_{dreg} and U_d is then given by

$$\frac{U_d}{F_{dreg}} = \frac{1}{D\left(1 + \left(\frac{2M}{D}\Delta\right)^2\right)} \frac{1 + \frac{2M}{D}s}{1 + 2 \times \frac{2M}{D\left(1+\left(\frac{2M}{D}\Delta\right)^2\right)}s + \left(\frac{2M}{D\sqrt{1+\left(\frac{2M}{D}\Delta\right)^2}}\right)s^2}. \tag{4.14}$$

Therefore, compared to that when the decoupling is perfect, i.e. $\frac{U_d}{F_d} = \frac{1}{D}\frac{1}{1+\frac{2M}{D}s}$, when the decoupling is imperfect, the transfer function of Eq. (4.14) is characterized by a:

- zero at $\frac{2M}{D}$,
- second-order denominator $1 + 2 \times \frac{2M}{D\left(1+\left(\frac{2M}{D}\Delta\right)^2\right)}s + \left(\frac{2M}{D\sqrt{1+\left(\frac{2M}{D}\Delta\right)^2}}\right)s^2$,
- reduced gain, by a factor $1 + \left(\frac{2M}{D}\Delta\right)^2$.

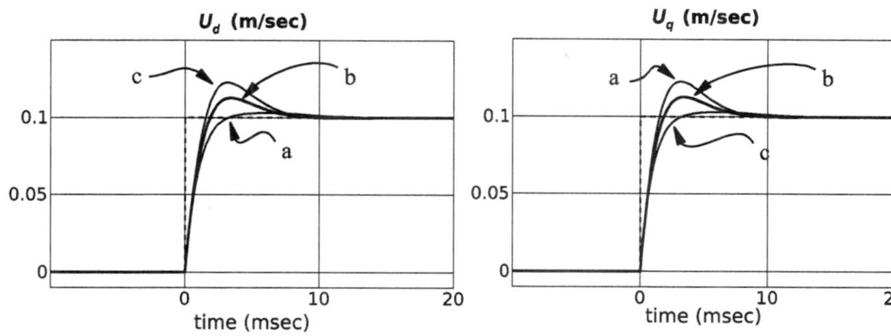

Figure 4.15 Transient response of U_d and U_q with the new controllers' gains calculated with $\Delta = 200$, for different estimation error of $\delta\omega$: (A) error of 200 rad/s, (B) no error, and (C) error of -200 rad/s.

In case of a wrong estimate of $\delta\omega$, we notice that a strong coupling between U_d and U_q still remains, as highlighted by the first ellipse in Fig. 4.14. To cancel out the effect of U_d on U_q, the *PI* controller in axis q has to adjust V_q to reject the disturbance, which takes time, approximately 25 ms in our example.

More importantly, the transient response at $t = 0$ is modified compared to the response of Fig. 4.12: it takes longer to reach the reference set point, and an overshoot is found on the response of U_d – see the second ellipse in Fig. 4.14. If we compare the response with that found in the example of Fig. 1.41, page 38, it seems that the dominant pole of the transfer function in axes d and q are overestimated. Actually, not only the dominant pole is faster, but the gain is smaller. To reduce the effect of the bad decoupling, it is possible to anticipate the error of resonance pulsation estimation, and to modify the correctors' gain accordingly. For instance, if we make the assumption that the error is smaller than a quantity named Δ, a rule of thumb consists in replacing D by

$$D = D\left(1 + \left(\frac{2M}{D}\Delta\right)^2\right) \tag{4.15}$$

in Eq. (4.9). The results obtained with these new settings are given in Fig. 4.15, with $\Delta = 200$ rad/s.

It is then possible to control the transducer at a different pulsation than the resonance one. However, it is difficult to maintain the response time when $\delta\omega$ is not perfectly estimated. If no extreme dynamic performances are required, this control is sufficient. In other cases, it is necessary to in-

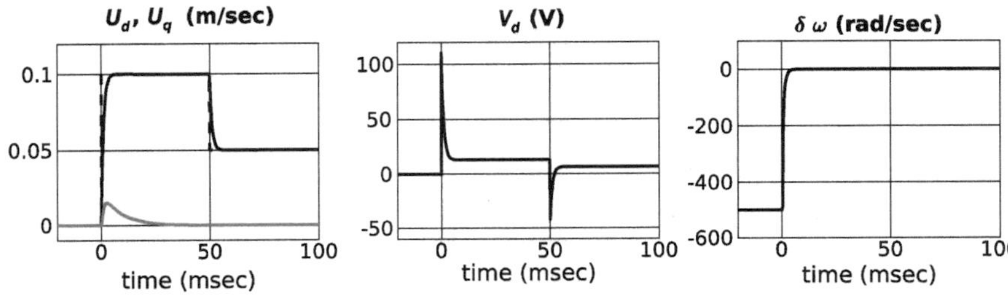

Figure 4.16 Transient response of U_d, U_q, V_d and $\delta\omega$ when simulating the model in the rotating reference frame.

clude a resonance frequency observer, which will be detailed in Sect. 4.4, or to track the resonance frequency as presented in the next section.

4.3.2 Simulation of the frequency tracking of a Langevin transducer

The frequency tracking is obtained easily by implementing the two already presented correctors in axes d and q. Their design is the same, and we show in Fig. 4.16 the evolution of U_d, U_q, $\delta\omega$ and V_d, when $\delta\omega$ is set to -500 rad/s at the beginning of the simulation.

It can be seen in Fig. 4.16 that after a few milliseconds, U_q reaches zero, meaning that the resonance pulsation is reached. The response time of U_d is 3 ms, and V_d is automatically adjusted by the corrector in order to obtain this response time.

These simulations validate the controllers, but they do not fully confirm the control scheme proposed in Fig. 4.9. We implement then this control by simulation using the lumped parameters model instead of the model in the rotating reference frame, and we take into account a bias in the estimation of the resonance pulsation, with $\omega_i = \omega_0 - 500$ rad/s. The simulation results are presented in Fig. 4.17.

Apparently, the vibration speed amplitude control is efficient. However, if we have a closer look at the variables in the rotating reference frame, we notice that the system is not operated at resonance all the time, as presented in Fig. 4.18.

Indeed, we noticed in Fig. 4.18 that at $t = 50$ ms, when U_d falls to 0.5 m/s, U_q increases, showing that the transducer does not operate at resonance (see ellipse 1). Indeed, the value of $\delta\omega$ suddenly increases (ellipse 2) while in this simulation run, the value of ω_0, the operating pulsation, should not change.

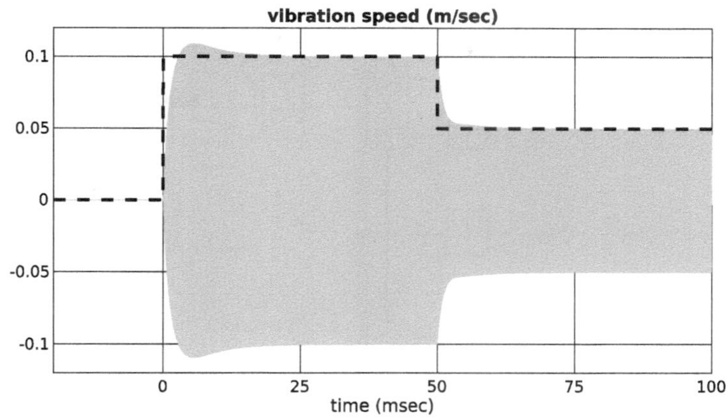

Figure 4.17 Vibration speed envelope (gray) and the amplitude reference (dashed) with the lumped parameters model and the control in the rotating reference frame.

To understand why this behavior occurs, we have to analyze Fig. 4.9 on page 144. In steady state, when U_d is constant, the value of $\delta\omega$ corresponds to the difference between ω_i and ω_0, so that the transducer operates at its resonant frequency. Therefore, the output of C_q is not null. When U_d changes – because U_{dref} has changed, for instance – $\delta\omega$ also changes instantaneously because the output of C_q is divided by U_d: the operating pulsation is then shifted. This error will be corrected after some time, by the loop corrector C_q which will converge to a new value. But at the end, the initial value of $\delta\omega$ is found, since neither ω_0 nor ω_i have changed.

This behavior is not optimal and it has to be changed. One possibility is to control $\frac{U_q}{U_d}$ instead of U_q. Indeed, at steady state, $U_q = 0$: if U_d changes, $\frac{U_q}{U_d}$ remains at 0, so the corrector C_q does not change its output, and the aforementioned drawback is avoided. Therefore, the control scheme of Fig. 4.9 is revised into that of Fig. 4.19.

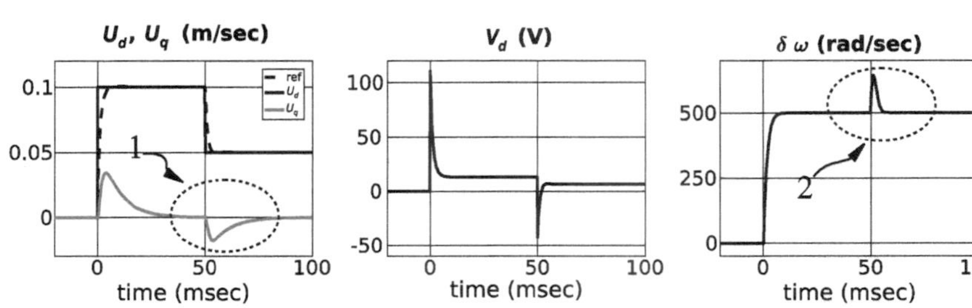

Figure 4.18 Transient response of U_d, U_q, V_d and $\delta\omega$ with the lumped parameters model.

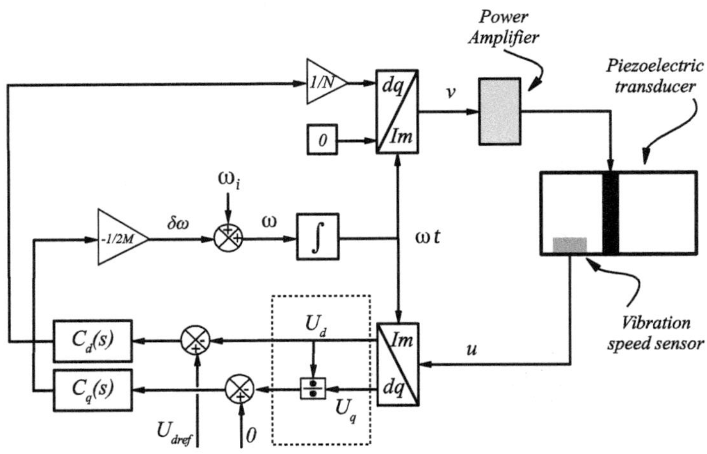

Figure 4.19 Modified control scheme of the vibration in the rotating reference frame with frequency tracking; the modification is highlighted with a rectangle.

Fig. 4.20 presents the simulation results obtained with this new control scheme. We notice that $\delta\omega$ reaches its steady state in 3 ms as expected, and is not affected by the transient of U_d at $t = 50$ ms. It should be emphasized here that the parameters of controller C_q have not been changed.

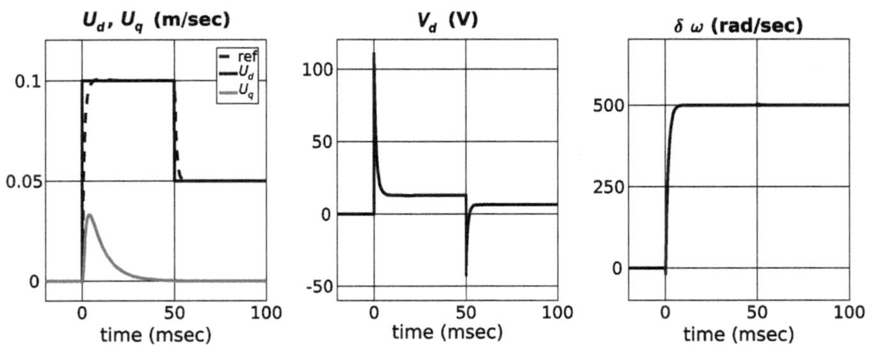

Figure 4.20 Transient response of U_d, U_q, V_d and $\delta\omega$ with the modified control scheme.

We have shown that this control can reach the resonance frequency of the system while vibrating at a given velocity amplitude. But what happens when the environment of the system (like the temperature) or the acoustical load changes its resonance behavior? This is tested in the next section of this chapter.

4.3.2.1 Robustness of the frequency tracking with respect to variation of the load damping

We consider that the load behaves like a damping and Eqs. (4.10) are valid:

$$F_{rd} = \tilde{D}U_d \quad \text{and} \quad F_{rq} = \tilde{D}U_q. \tag{4.16}$$

Here also we remove the compensations of F_{rd} and F_{rq} in the storage inversion block. At $t = 0$, the reference of the vibration velocity is set to $U_{dref} = 0.1$ m/s, and at $t = 50$ ms, the load damping is set, and we have $\tilde{D} = 0$ if $t < 50$ ms, while $\tilde{D} = D$ if $t > 50$ ms. The results are given in Fig. 4.21.

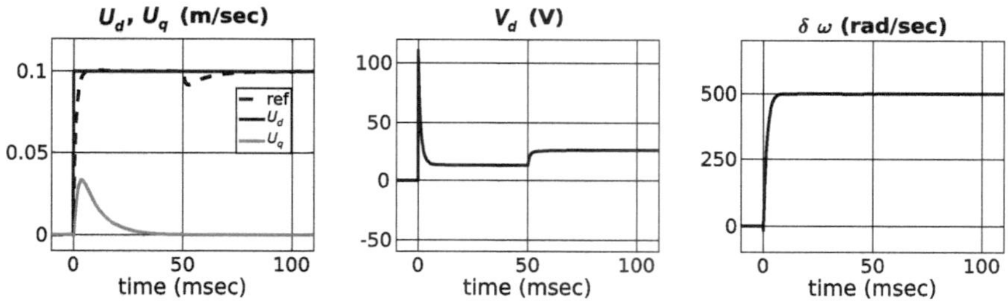

Figure 4.21 Effect of a variation of the load damping on the transient response of U_d, U_q, V_d and $\delta\omega$.

Fig. 4.21 shows that at $t = 0$, the voltage increases in order to reduce the rise time of U_d and then is reduced in order to maintain it at its reference. When the perturbation occurs from the load, the voltage is not immediately increased, therefore U_d shows a slight drop, the time for C_d to compensate the perturbation by increasing its output. In axis q, U_q is not affected by the perturbation from the load: since $U_q = 0$, we have $F_{rq} = 0$, too. Therefore, the value of $\delta\omega$ is not changed.

4.3.2.2 Robustness of the frequency tracking with respect to variation of the load equivalent stiffness

When the load can be modeled by a pure equivalent stiffness, the perturbation force f_r is written as $f_r = \tilde{K}w$. Following Eq. (3.35) on page 110, we have

$$F_{rd} = \tilde{K}\left(\frac{U_q}{\omega} + \frac{\dot{U}_d}{\omega^2}\right) \quad \text{and} \quad F_{rq} = \tilde{K}\left(-\frac{U_d}{\omega} + \frac{\dot{U}_q}{\omega^2}\right). \tag{4.17}$$

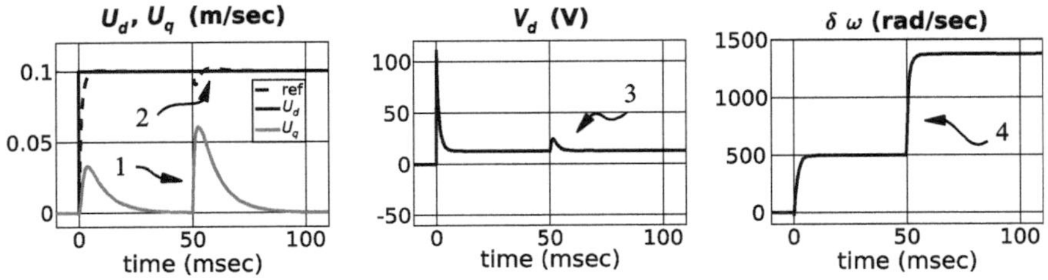

Figure 4.22 Effect of a variation of the load stiffness on the transient response of U_d, U_q, V_d and $\delta\omega$.

Hence the load produces a perturbation in axes d and q, and there is a coupling between them. In fact, the effect of the load is equivalent to a resonance frequency shift, as explained in Sect. 3.3.2. We verify in Fig. 4.22 whether this perturbation can be rejected and the frequency automatically adjusted in order to track the resonance. For this the reference of the vibration velocity is set at $t = 0$ to $U_d = 0.1$ m/s, and then, at $t = 50$ ms, $\tilde{K}$ is set to $-1\%K$ which changes the resonance frequency from 27878 to 28017 Hz (the difference is 139 Hz).

It can be seen from Fig. 4.22 that when the load stiffness is increased, U_q instantaneously increases, as it was expected from Eq. (4.17) (label 1 in Fig. 4.22). Due to the coupling between U_d and U_q, U_d varies because of the variation of U_q (label 2), and therefore, the corrector C_d compensates for the effect of U_d by increasing its output (label 3). During this time, the corrector C_q increases the value of the operating pulsation in order to maintain $U_q = 0$, from 500 to 1370 rad/s – corresponding to the 139 Hz difference (which is expected from the calculation of the resonance frequency before and after the change of $\tilde{K}$) – in 3 ms.

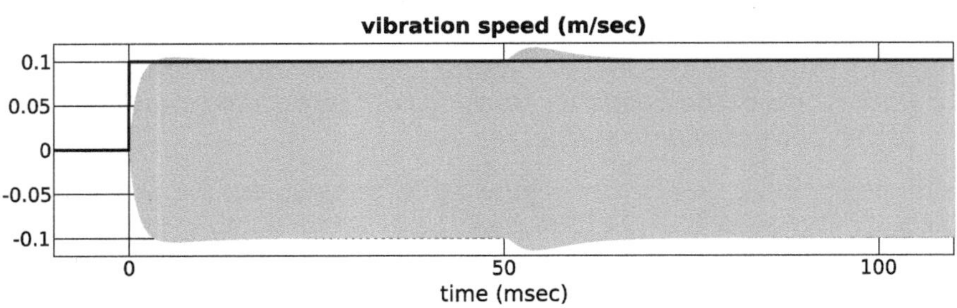

Figure 4.23 Effect of a variation of the load stiffness on the transient response of U_d, U_q, V_d and $\delta\omega$.

In spite of this control, the amplitude of the vibration velocity slightly deviates from its reference at $t = 50$ ms (Fig. 4.23). This is due to the rise of U_q[1]; however, the deviation is limited to 15%, which is reasonable.

4.4. Observers

In some applications, it can be very important to have an information about the force produced by the environment on the transducer f_r, while the vibration velocity u is controlled. For example, if the device is used in a water treatment application, like medicine removal (Méndez-Arriaga et al., 2008), $f_r = 0$ means that the transducer has no liquid to treat, and the vibration should stop. If the device is an ultrasonic motor, the force is correlated with the torque to the load, and knowing f_r can be used to control the tightening force of a claw (Giraud and Semail, 2006), without torque sensor.

It is, however, very difficult to measure $f_r(t)$. Indeed, this force is sinusoidal, at the same pulsation as the vibration, and most of the time, f_r actually is distributed over a surface. Therefore, we will use an *observer* for f_r.

Observers are used when a state variable cannot be measured; we use then the measurement of other variables to construct – thanks to a model – the missing variable. To start with an observer, a model is thus required.

Since the device is controlled in the rotating reference, the model has been established in Chap. 3, and we recall here Eqs. (3.43) on page 112:

$$\begin{aligned} 2M\dot{U}_d + DU_d &= F_d - F_{rd}, \\ 2M\dot{U}_q + DU_q &= F_q - F_{rq}. \end{aligned} \tag{4.18}$$

However, if we assume that F_d is perfectly controlled and U_d is measured, then calculating $\tilde{F}_{rd}$, the observed value of F_{rd}, directly from (4.18) is not a good option. Indeed,

$$\tilde{F}_{rd} = F_{dreg} - 2M\dot{U}_d + DU_d \tag{4.19}$$

leads to calculating the first derivative of U_d. This calculation is difficult, because derivative increases the signal-to-noise ratio. Therefore, we will use a closed-loop observer instead.

A closed-loop observer consists in replicating the model of the system, and to control it from the observed variable (here $\tilde{F}_{rd}$) (Bouscayrol et al.,

[1] The amplitude of the vibration velocity u is equal to $U = \sqrt{U_d^2 + U_q^2}$.

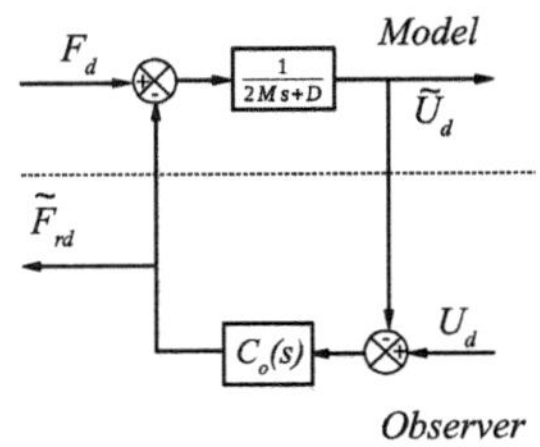

Figure 4.24 A closed-loop observer.

2009). The reference is not U_{dref} but rather U_d. This way we look for the input F_{rd} which gives, for the same force F_d, the same vibration velocity amplitude U_d. This is depicted in Fig. 4.24, where the observer's controller is denoted by C_o.

It should be emphasized here that C_o relies on the same *FTBO* as the controllers C of axes d and q; therefore, C_o and C can be equivalent.

We consider a load composed of a damping $\tilde{D}$ and a stiffness $\tilde{K}$ and we write:

$$F_{rd} = \tilde{D}U_d + \tilde{K}\left(\frac{U_q}{\omega} + \frac{\dot{U}_d}{\omega^2}\right) \quad \text{and} \quad F_{rq} = \tilde{D}U_q + \tilde{K}\left(-\frac{U_d}{\omega} + \frac{\dot{U}_q}{\omega^2}\right). \tag{4.20}$$

For the purpose of illustration, the observers in axes d and q are implemented in the control at fixed frequency of a transducer detailed in Sect. 4.3.1. The observers' controllers are equal to the vibration velocities controllers, so they have the same dynamic (response time $t_r = 3$ ms).

At $t = 50$ ms the references $U_{dref} = 0.1$ and $U_{qref} = 0.1$ are set, and at $t = 100$ ms, a step variation of $\tilde{K} = 0.25\%K = 19.1 \times 10^6$ N/m and $\tilde{D} = D = 58$ Ns/m are imposed on the load. The results of the observers are presented in Fig. 4.25.

The result show that the observers are useful to estimate the forces acting on the transducer from the load. For these estimations, we can also deduce the equivalent acoustic impedance of the load at the working frequency. For instance, we can calculate $\tilde{D}$ and $\tilde{K}$ by

$$\tilde{D} = \frac{F_{rd}U_d + F_{rq}U_q}{U_d^2 + U_q^2} \quad \text{and} \quad \frac{\tilde{K}}{\omega} = \frac{F_{rd}U_q - F_{rq}U_d}{U_d^2 + U_q^2}. \tag{4.21}$$

However, the results are obtained for the ideal case when $\delta\omega$ is perfectly known. Fig. 4.26 depicts the result of the observers when an error on $\delta\omega$ is made ($\delta\omega = 0.5\%\omega_0$) which appears at $t = 0.1$ s.

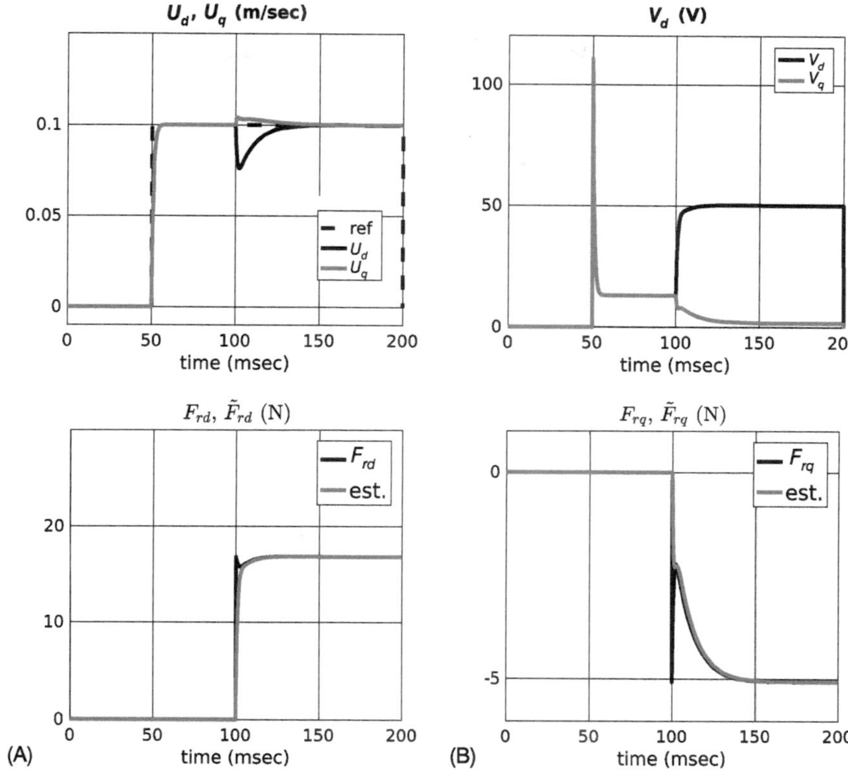

Figure 4.25 Results of the reaction force observer at fixed frequency: (A) U_d, U_q, V_d and V_q, (B) comparison between F_{ri} and $\tilde{F}_{ri}$, $i \in \{d, q\}$.

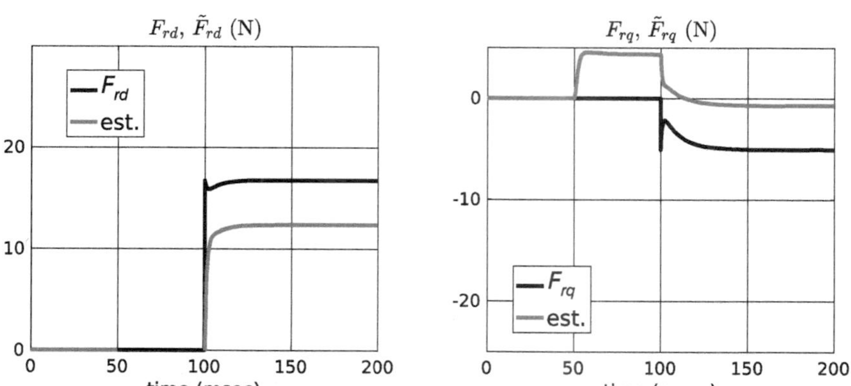

Figure 4.26 The observer at a fixed frequency, introducing an error on the estimated value of $\delta\omega$.

Compared with the ideal case, the observation of F_{rd} and F_{rq} now have a static error. In fact, since $\delta\omega$ is not known, the couplings between axes d and q are not compensated for. The forces F_d and F_q differ then from their references, leading to the error in the observations of F_{rd} and F_{rq}. In this example, the estimates of $\tilde{K}$ and $\tilde{D}$ expressed in Eq. (4.21) would give $\tilde{D} = 58$ Ns/m and $\tilde{K} = 1\%K$: $\tilde{D}$ is still well estimated, but not $\tilde{K}$.

Hence, for a good observation of the reaction forces, it is necessary to rely on an accurate model of the system. To limit discrepancies, it is therefore recommended to identify the system at no load as frequently as possible.

CHAPTER FIVE

Applications

Contents

5.1. Introduction 161
5.2. Control of a Langevin transducer 162
- 5.2.1 Parameters identification and corrector design 162
- 5.2.2 Experimental results 164

5.3. Control of a traveling wave in a finite beam 165
- 5.3.1 System description and controllers design 165
- 5.3.2 Experimental results 167

5.4. Control of a traveling wave ultrasonic motor 170
- 5.4.1 Objectives of the control 170
- 5.4.2 Identification of the stator parameters 172
- 5.4.3 Experimental results 174

Abstract

This chapter presents experimental results obtained on specific applications. In the first application, a Langevin transducer is controlled in amplitude and frequency, and we will see the effect of nonlinearities on the closed-loop control. In the second application, we show a traveling wave obtained on a finite beam; this example can be used as a linear motor. Then, we present the vector control of a traveling wave ultrasonic motor. Here also, we show the behavior of the system with regards to the nonlinearities of the stator/rotor contact mechanics.

Keywords

Langevin transducer control, traveling wave, transfer function, controller, mode coupling, parameters identification

5.1. Introduction

Realizing a practical framework for closed-loop control of a piezoelectric transducer is often a matter of experience. Capitalizing on the results from one application to another is rare in the literature, and the engineers redo the work of the controller's design. In this chapter, the methodology presented in Chap. 4 is applied in typical applications which are a Langevin transducer, traveling wave in a linear beam, and control of a traveling wave ultrasonic motor. These applications need high performances in terms of response time, synchronization, and robustness to

Piezoelectric Actuators: Vector Control Method
https://doi.org/10.1016/B978-0-12-814186-1.00009-0

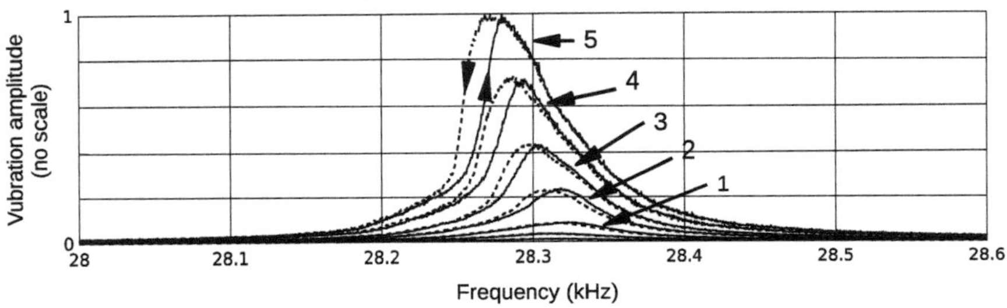

Figure 5.1 Frequency response of a Langevin transducer at different levels of voltage ($V = 60$, 80, 100, 120 and 140 volts peak-to-peak, respectively).

the nonlinearities. We will see that the vector control method allows us to dynamically and precisely control vibration amplitude, and that the controller's design presented in Chap. 4 can be applied in practice.

5.2. Control of a Langevin transducer

5.2.1 Parameters identification and corrector design

We consider a Langevin transducer (FBL28252, Fuji electroceramics, Japan). The power supply comes from a linear amplifier (HSA4051, NFcorp., Japan), and the vector control is embedded into a microcontroller (STM32F4, ST Microelectronics, France). This device can be used for several applications, like ultrasonic cleaning, welding or cutting in a medical application. In this example, we want a control of the vibration amplitude to a reference value, with a required response time of $t_r = 30$ ms, and a resonance frequency tracking in order to reduce the power consumption. The open-loop characteristics of the device, like the frequency response of the vibration for several voltages as depicted in Fig. 5.1, show a nonlinear behavior with pull-out (see page 32). Since the response shows a small bandwidth, it is then necessary to track the resonance frequency in order to operate at a lower voltage amplitude value.

To determine the parameters of the model, we use the representation of the operating points in the rotating reference frame as depicted in Fig. 5.3. Indeed, as presented in Sect. 3.3.2, when we draw U_d as a function of U_q with $V_d = V$ and $V_q = 0$, the points follow a circle of radius $\frac{NV}{2D}$, while the pulsation at the maximum and minimum of U_q gives the bandwidth (points P_1 and P_2 in Fig. 5.3). To avoid the effect of the nonlinearities, we maintain V at a small value.

How can we estimate N?

In a piezoelectric transducer, the force factor N describes the energy conversion inside the piezoelectric material. Its value can be calculated, provided that we exactly know the material's parameters, the dimension of the device and the deformation mode shapes. Estimation from measurement however possible, at the expense of equipment which is not always found in laboratories.

A method consists in measuring the motional current flowing to the device $i_m(t)$, and comparing it to the vibration speed $u(t)$. Indeed, according to Eq. (2.135), we have $i_m = N \times u$. For this, we use a laser interferometer to measure the vibration speed $u(t)$ (like the IVS-500 from Polytec, Germany), a current probe to measure the current $i(t)$ into the device (like the TCP202A from Tektronix, USA), a voltage probe to measure the voltage $v(t)$ applied to the device (like the P2220 from Tektronix), and an oscilloscope to record the data (like TDS 2014C from Tektronix). The experimental setup is given in Fig. 5.2.

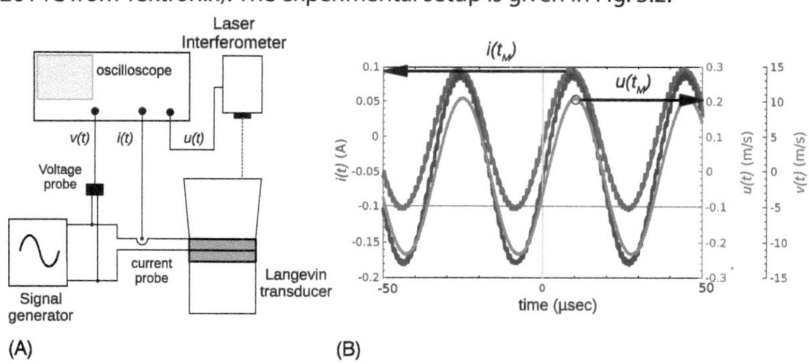

Figure 5.2 Experimental setup for the identification of N (A) and identification of N for an experimental record (B); t_M has been chosen at a maximum of voltage. We find $i(t_M) = 0.095$ A, $u(t_M) = 0.2$ m/s, leading to $N = 0.475$ N/V.

Unfortunately, the current i_m cannot be measured directly: it is an internal variable of the model. However, at some times denoted t_M, $i(t_M) = i_m(t_M)$. Indeed, the electric model of a piezoelectric transducer shows that $i(t) = i_m(t) + C_0 \frac{dv}{dt}$. Hence, at time t_M where $\frac{dv}{dt} = 0$, $i(t_M) = i_m(t_M)$. In theory, it is easy to identify t_M: this is the time when the voltage is at a maximum or minimum.

If we denote $I_M = i(t_M)$ and $U_M = u(t_M)$, the values of the current and the vibration speed at $t = t_M$, the estimated value of N is equal to

$$N = \frac{I_M}{U_M}. \tag{5.1}$$

An identification is presented in Fig. 5.2 (B), which yields $N = 0.475$ N/V.

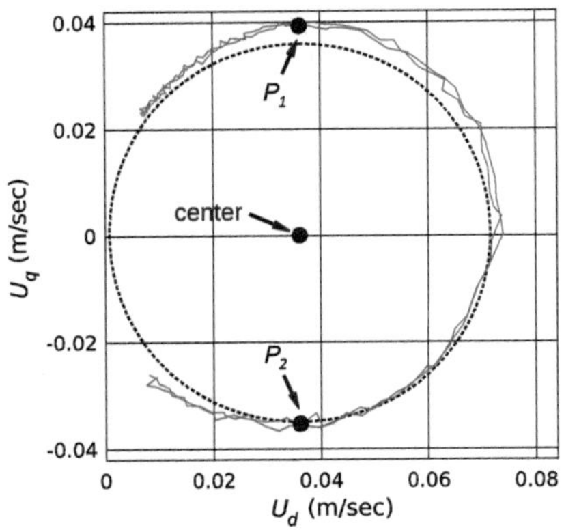

Radius of the circle	$\frac{NV}{2D}$	0.037 m/s
bandwidth	ω_{BP}	1162 rad/s
Gain	$\frac{N}{D}$	0.74 mm/s/V
time constant $\frac{2M}{D}$	$\frac{2}{\omega_{BP}}$	1.7 ms

Figure 5.3 Steady state operation in the rotating reference frame, and the identified parameters.

From the identified value of the radius of the circle ($R = \frac{NV}{2D}$) and the bandwidth of the system (equal to $\omega_{BP} = \omega(P_2) - \omega(P_1)$), we can identify the parameters of the open-loop transfer function. The parameter values are given in Fig. 5.3. However, this identification process doesn't allow us to know the value of the force factor N. To achieve the control of the vibration speed and track the resonance frequency, we can adopt $N = 1$ without lack of generality. However, the forces calculated with this value will not be equal to the internal forces in the piezoelectric system.

Since the vibration amplitude control and resonance frequency tracking are both necessary in this example, we adopt the control scheme detailed in Fig. 4.19, on page 154. The parameters of the two controllers, C_d and C_q, are equal, and we obtain $K_p = 115$ and $K_i = 0.195$. The value of V_q is set to 0.

5.2.2 Experimental results

Fig. 5.4 (A) shows the experimental result of the frequency tracking. In this try, U_{dref} is set to 50 mm/s, corresponding to a vibration amplitude of 0.29 μm peak while $U_{qref} = 0$; the controllers are switched on at $t = 0$. Observe that for $t < 0$, we have $U_q > 0$, meaning that the operating frequency is smaller than that of the resonance. At $t = 0$, the frequency controller on axis q increases the operating pulsation until $U_q = 0$. As a result, the operating pulsation gets closer to that of the resonance, and the coupling between the axes d and q decreases: the voltage amplitude – which is equal to V_d

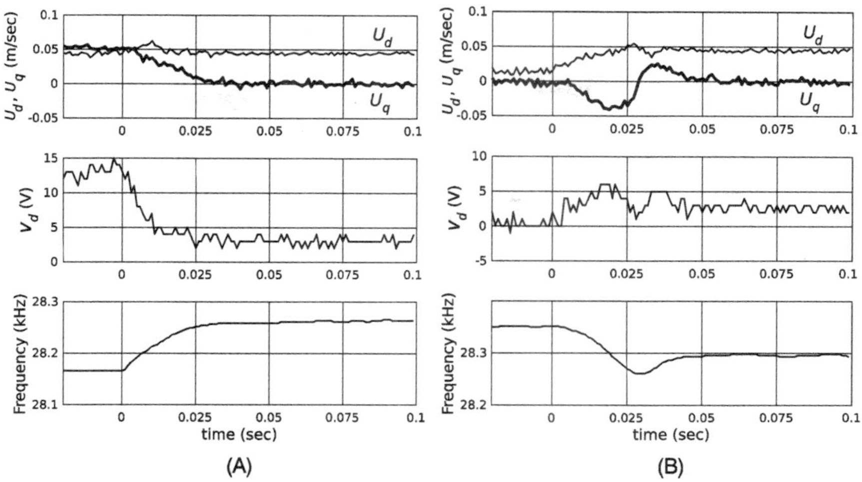

Figure 5.4 Experimental result of the closed loop control of a Langevin transducer: (A) frequency tracking, (B) vibration velocity control.

in this case – is decreased by the controller C_d, leading to a more efficient operation.

When we want to dynamically change the vibration velocity U_d, we need to increase voltage V_d. However, due to internal nonlinearities, an increase in vibration amplitude is accompanied by a decrease in the resonant frequency. Therefore, a dynamic tracking of the resonant frequency is necessary in this application. This is automatically achieved by the control structure as depicted in Fig. 5.4. When the reference U_d increases from 10 to 50 mm/s, the action of C_d increases the voltage, and the vibration amplitude increases. At the same time, the resonance frequency shifts and decreases due to the nonlinearity of the transducer; therefore, U_q becomes negative. The controller C_q decreases the working frequency automatically, in order to have $U_q = 0$ again.

5.3. Control of a traveling wave in a finite beam

5.3.1 System description and controllers design

We consider a finite beam ($493 \times 6 \times 6$ mm^3) made of aluminum. The forces acting on the beam are produced by two piezoelectric transducers (NAC2003H06 from CTS, USA). To obtain a traveling wave, we have to energize two successive modes on the beam. Mode *a* is the 9th bending

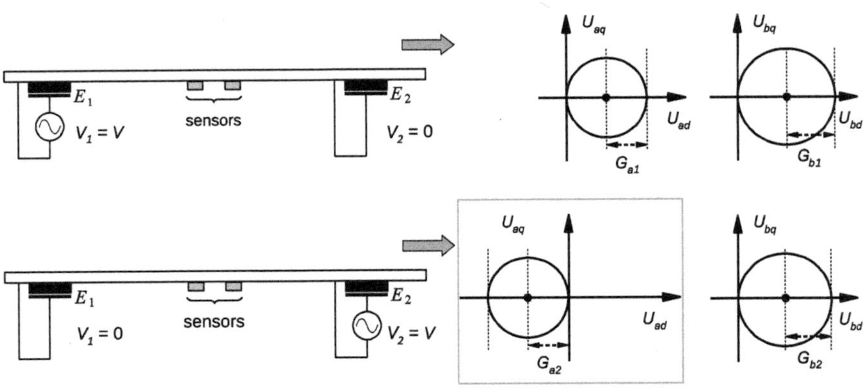

Figure 5.5 Identification of the transfer functions and Φ for a traveling wave excitation; the circles are used to determine the values of the gains G_{ik} and τ_{ik}. Gain G_{a2} is negative because the circle is located in the mid-plane $U_{ad} < 0$ (box).

mode and has a resonance frequency of 17900 Hz. Mode b is the 10th bending mode, with a resonance frequency of 20200 Hz. We want a closed-loop response time of 100 ms.

The identification of the system requires identifying:

- the transfer functions of the two modes, so 4 parameters in total, and
- the matrix Φ.

To do so, we should identify the deformation mode shape of each mode, and identify the precise location of the actuators to finally estimate the values of Φ. This is, however, not practical as it requires expensive measuring devices, and so an indirect method will be preferred. This method consists in drawing the frequency responses of the two modes, when one actuator voltage is set to 0, while the other is set to an arbitrary value. For this, the frequency range is limited around the resonance frequency of the studied mode. By this procedure, we identify 4 gains, denoted by G_{ik}, and 4 response times τ_{ik} where $i \in \{a, b\}$ and $k \in \{1, 2\}$. For example, G_{a1} represents the identified gain when $V_2 = 0$ for mode a. This method is described in Fig. 5.5.

As mentioned in the previous section, this method doesn't help identify N but it helps identify Φ, G_{ik} ,and τ_i which can be written as:

$$G_i = \frac{N}{D_i} = \sqrt{G_{i1}^2 + G_{i2}^2}, \quad \Phi_i(x_k) = \frac{G_{ik}}{G_i}, \quad \tau_i = \frac{2M_i}{D_i}. \tag{5.2}$$

If we consider that $N = 1$, without lack of generality, the identified values of the transfer functions gains are summarized in Table 5.1.

Table 5.1 Identified values of the transfer function gains and matrix Φ.

G_{a1}	G_{a2}	G_{b1}	G_{b2}
0.0190	0.0217	0.0159	−0.017
G_a		G_b	
0.0288		0.0230	
$\Phi_a(x_1)$	$\Phi_a(x_2)$	$\Phi_b(x_1)$	$\Phi_b(x_2)$
0.659	0.752	0.683	−0.730

In theory, the time constants τ_{i1} and τ_{i2} should be equal, as they refer to the same transfer function. In practice, the identified values can be different, due to noise in the measured signal or nonlinearities in the system, for instance, but should not differ by a large value. Therefore, we will calculate the average values from the two measurements as

$$\tau_i = \frac{\tau_{i1} + \tau_{i2}}{2}. \tag{5.3}$$

From the measurements, we identified that the two time constants are the same with

$$\tau_a = \tau_b = 1.9 \text{ ms}. \tag{5.4}$$

We have 4 loop correctors, 2 for each mode and for axes d and q. They are denoted by $C_{ak}(s) = K_{pa} + \frac{K_{ia}}{s}$ and $C_{bk}(s) = K_{pb} + \frac{K_{ib}}{s}$, respectively, and with $k \in \{d, q\}$. The gains are different, but they are the same for one mode. They are calculated using the same procedure, detailed on page 37. We find: $K_{pa} = 1979$, $K_{pb} = 2478$, $K_{ia} = 3.76$, and $K_{ib} = 4.71$.

5.3.2 Experimental results

The closed-loop control is first tested on each mode separately. For this, a reference step is programmed at $t = 0$ for U_{aq} while U_{ad} remains equal to 0. During this step, the voltage on mode b is fixed to 0. The system's response is depicted in Fig. 5.6. As we can see, the response time is equal to 0.1 s. Moreover, mode b is not affected by the response, showing that there is no coupling between modes a and b.

Once the control loops for the two modes are checked, a traveling wave can be obtained by setting references for U_a and U_b. In this example, we have $U_{ad} = U_{bq} = 0$ and $U_{aq} = U_{bd} = 7$ mm/s, as presented in Fig. 5.7.

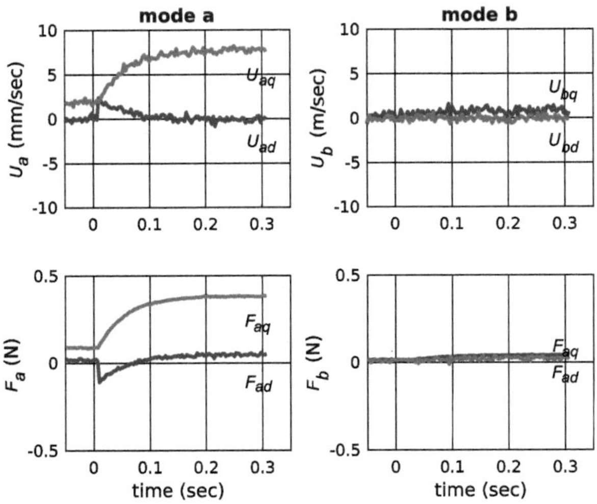

Figure 5.6 Closed-loop response of mode *a* to a step variation of the reference.

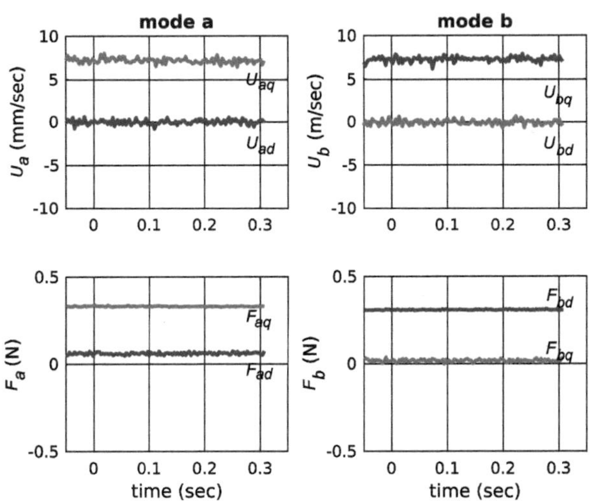

Figure 5.7 Vibration speeds of the 2 modes in order to obtain a traveling wave, and the modal forces, as functions of time.

We present in Fig. 5.8 the vectors of the vibration speeds, modal forces and voltages set on the actuators in the rotating reference frame. As it can be seen, the vibration speeds U_a and U_b are orthogonal in the rotating reference frame. However, to obtain this result, the two modal forces are not orthogonal. This is due to the fact that the two modes don't have the

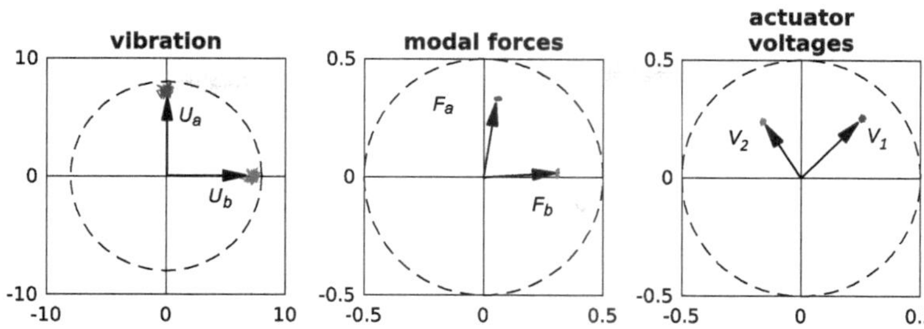

Figure 5.8 Vibration speeds, modal forces and voltages on the actuators in the rotating reference frame for a traveling wave.

same open loop gain, and also because the working frequency is between the two modes. The controllers adapt the modal forces in order to obtain the wanted vibration speeds. Finally, the voltage supplied to the actuators is calculated using the inverse of matrix Φ_{ab}. Here also, the voltages are not orthogonal, and don't have the same length (amplitude). This results from the position of the actuators on the beam.

For some reason, changing the direction of the traveling wave is necessary to change the direction of a vehicle when the traveling wave is used in a motor, for instance. To do so, the voltage phase and amplitude need to be changed. With the closed-loop control, we just change the references for the vibration speed amplitude of one mode. In our example, we set $U_{dref} = -U_{dref}$. The experimental results are presented in Fig. 5.9.

The control loops succeed in reversing the direction of the traveling wave in 0.1 s. It is often admitted that reversing the direction of a traveling wave is achieved simply by switching the voltages V_1 and V_2 (*i.e.*, $V_2 = V_1$ and $V_1 = V_2$). However, this is not the case here: if we look at the actuators' voltages, we see the head of V_1 before the direction change is not located at the head of vector V_2 after the direction change, so they are not equal. This is due to the fact that the two actuators don't have the same position on the beam. In fact, changing the direction leads to having $F_b = -F_b$. But when decoupling the modes (through Φ_{ab}^{-1}), the two voltages V_1 and V_2 have totally different values. In theory, if the actuators are located symmetrically on the beam, reversing the direction is achieved by switching the voltages. However, in practice, this condition is difficult to obtain, and a closed-loop control gives better results.

Since the vibration speeds are controlled for each mode, the system becomes robust besides change in working frequency. For example, Fig. 5.10 shows the response of the system when the working frequency is changed

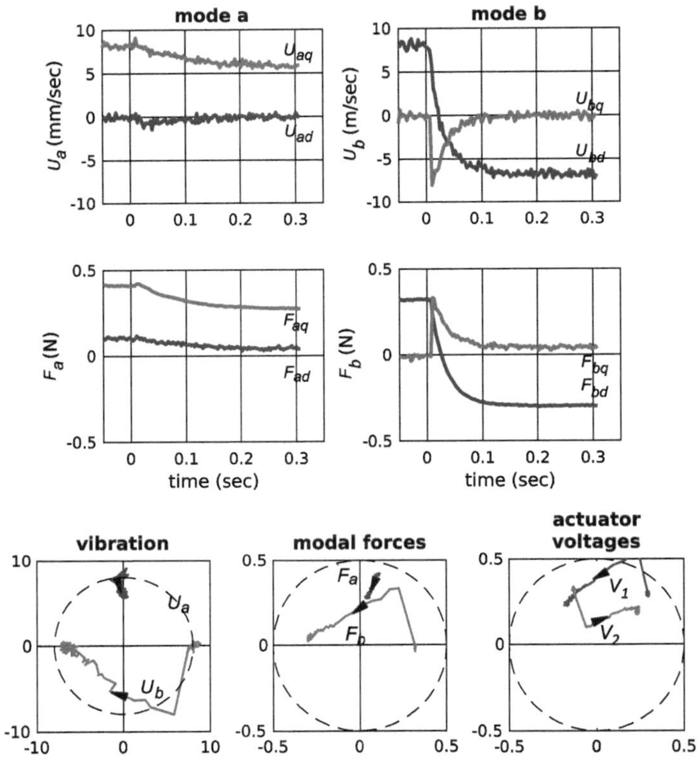

Figure 5.9 Response of the traveling wave to a direction change at $t = 0$; in time domain (top) and in the rotating reference frame (bottom).

from 18800 to 19200 Hz at $t = 0$. We remark that to maintain the vibration speeds to their reference, the forces have changed. This is due to the fact that the coupling terms between axes d and q for each mode are modified, because the values of $\delta\omega_a$ and $\delta\omega_b$ are changed. For instance, F_a is increased and F_b is decreased because when we increase the working frequency, we also get closer to mode b. Automatically, the change in the working frequency leads to updated values of the voltages V_1 and V_2.

5.4. Control of a traveling wave ultrasonic motor

5.4.1 Objectives of the control

The traveling wave ultrasonic motor (TWUM) uses two vibration modes in order to drive the rotor firmly pressed on the stator. Chapter 3 illus-

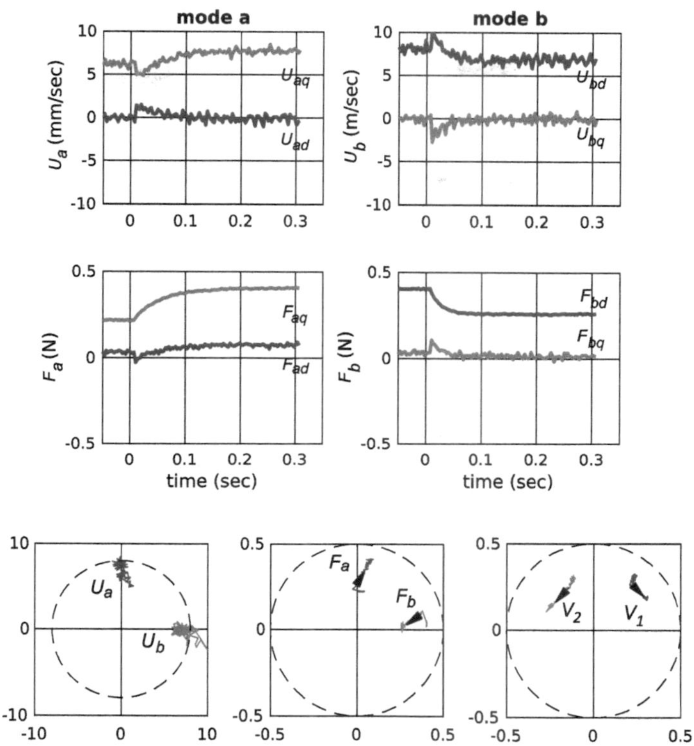

Figure 5.10 Response of the traveling wave to a working frequency change at $t = 0$; in time domain (top) and in the rotating reference frame (bottom).

trates the effect of this contact between the stator and rotor. Indeed, *the pull-out* phenomenon appears. Therefore, tracking the resonance frequency is tricky, because it is not stable, but depends on the vibration amplitude of the stator: increasing the vibration amplitude decreases the resonance frequency. Moreover, the torque produced by the motor damps the vibration of the rotor, and more voltage is needed when the torque increases.

Most of the TWUM controllers operate at constant voltage amplitude (like 200 V) but with a varying frequency. The control of the vibration amplitude is then achieved by changing the frequency. But due to the aforementioned characteristics of the motor, the manufacturers take a margin, and don't increase the vibration amplitude too much, which would take the motor closer to its unstable resonance point. This operating principle reduces the efficiency of the motor and its power amplifier.

In this section, we will use VCM to operate a TWUM. We will show how the voltage can be reduced, yet ensuring a stable and efficient operation of the motor.

5.4.2 Identification of the stator parameters

In this section, we study a USR60 from Shinsei (Japan). This motor has two phases, denoted 1 and 2. The operating frequency is between 40 and 44 kHz, for 150 rotation per minute and 1 Nm of torque as rated values. The operating voltage is 200 V.

Section 3.4.4.3 presents a simplified model of the TWUM for the specific case when the deformations of the two phases are shifted by $\frac{\pi}{2}$. Under such a condition, the model is equivalent to phase 1 only, which controls mode a. To ensure the assumptions of the model, the voltage on phase 2, denoted v_2, is deduced from v_1 by using the following relations:

$$\begin{aligned} V_{2q} &= -V_{1d}, \\ V_{2d} &= V_{1q}. \end{aligned} \tag{5.5}$$

Moreover, if the control ensures that $U_q = 0$, then axis d controls the vibration amplitude with the torque as a perturbation. For the sake of clarity, the two equations on each axis are recalled here:

$$\begin{aligned} DU_{ad} = F_{pd} - F'_T \rightarrow F_{pd} = DU_{ad} + F'_T, \\ 2M\delta\omega U_{ad} = F_{pq} + F_N \rightarrow F_{pq} = 2M\delta\omega U_{ad} - F_N. \end{aligned} \tag{5.6}$$

The strategy followed here consists then in:

- achieving a frequency tracking of the motor, in order to have $U_q = 0$;
- controlling the voltage to keep $U_d = U_{dref}$.

The three parameters necessary to design the control are D, M and N. However, we cannot apply the same methodology as described in Sect. 3.3, because the assumption $U_{aq} = 0$ is not guaranteed. Therefore, to estimate D, the method consists here in setting $V_{1q} = 0$ on an unloaded motor ($F'_T = 0$) and sweeping the frequency (from high to low) at several values of V_{1d}. We then measure the maximum of the vibration velocity, which corresponds to resonance ($U_{aq} = 0$). Eq. (5.6) gives

$$NV_{1d} = DU_{ad}. \tag{5.7}$$

Then, plotting V_{1d} as a function of U_{ad} gives a straight line, with slope equal to $\frac{D}{N}$. The result is presented in Fig. 5.11.

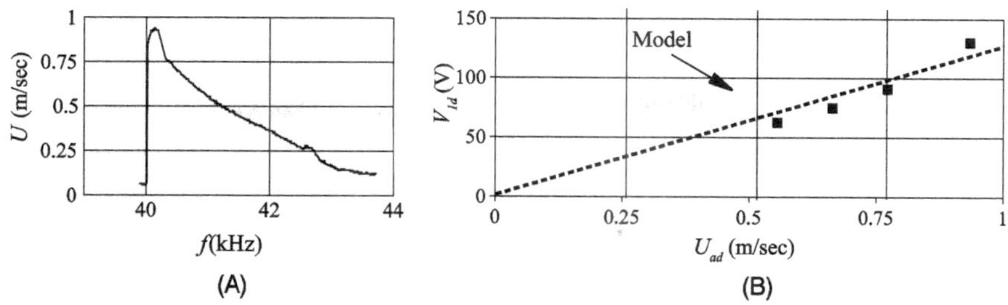

Figure 5.11 Frequency response of a TWUM for $V_{1d} = 130$ V (A), and voltage as a function of the maximum vibration velocity amplitude at resonance (B).

In practice, due to the fact that the resonance point is very unstable and thus hard to identify, the points are not aligned, contrary to the model. However, we can find an average value, which in our case is $\frac{N}{D} = 8.7$ mm/s/V. The force factor N is estimated to be $N = 0.32$ N/V, leading to $D = 37$ Ns/m.

To estimate M, we first choose a vibration velocity amplitude U_s and set $V_{1q} = 0$ and $V_{1d} = 130$ V. Pulsation ω is tuned in order to reach the required value of $U = U_s$. While maintaining ω constant, V_{1q} is decreased in order to have $U_{aq} = 0$, and V_{1d} is reduced to have $U_{ad} = U_s$. From that point, and still with $F'_T = 0$, we tune V_{1q} to reach $U_{aq} = 0$ for several values of ω. Hence, due to Eq. (5.6), the relation between V_{1q} and ω is linear, with a slope equal to $\frac{2MU_s}{N}$. (See Fig. 5.12.)

This identification leads to $M = 28$ g. Thanks to these parameters, the motor can be controlled in the rotating reference frame.

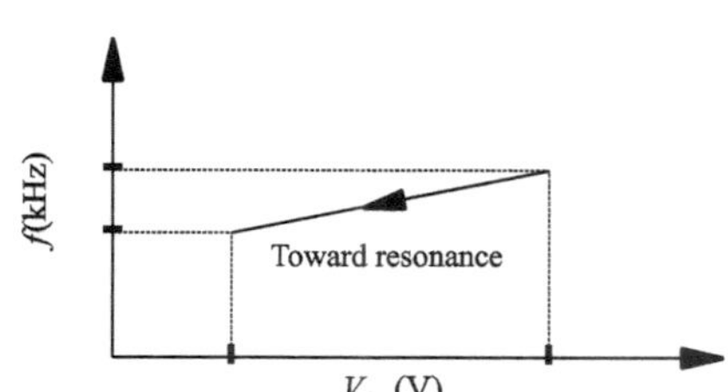

Figure 5.12 Relation between voltage V_{1q} and the frequency at constant vibration velocity amplitude.

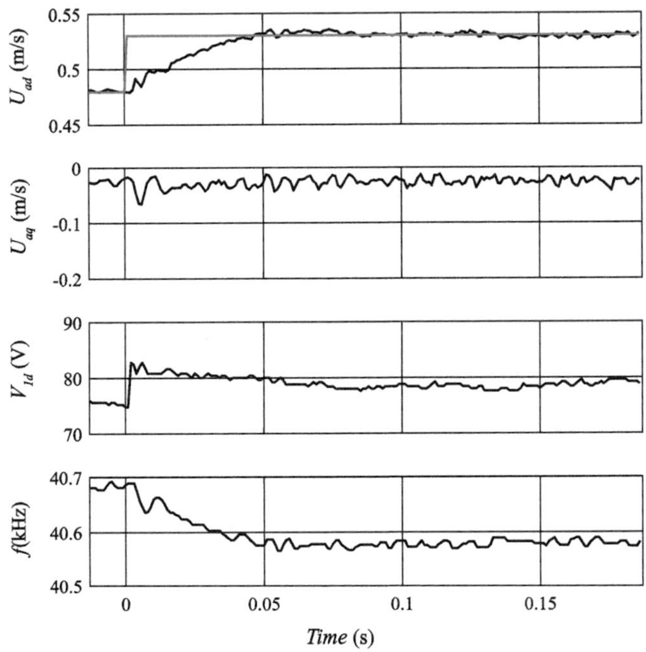

Figure 5.13 Transitory response of the TWUM to a step variation of U_{adref} from 0.48 to 0.53 m/s.

5.4.3 Experimental results

To operate the motor with the lowest voltage, we will achieve a resonance frequency tracking of the motor. For this, we will set $V_{1q} = 0$, and the pulsation ω will be tuned by the controller in order to have $U_{ad} = 0$. We adopt then the control scheme detailed in Fig. 4.19, on page 154, for one phase of the motor. The parameters of the two controllers, C_d and C_q, are equal, and we obtain $K_p = 3.7$ and $K_i = 0.0056$ for a response time at 5% equal to 45 m. The results of the transitory response of the TWUM controlled in the rotating reference frame are depicted in Fig. 5.13.

It can be seen that the vibration amplitude follows its reference; the response time of 45 ms is respected. To achieve this, voltage V_{1d} is increased; its value during the transient is higher than in steady state, in order to accelerate the stator's vibration response. And since the vibration amplitude is increased, the operating frequency has to change from 40.7 to 40.6 kHz in order to maintain $U_{aq} = 0$. This is automatically achieved thanks to the resonance frequency tracking controller in axis q.

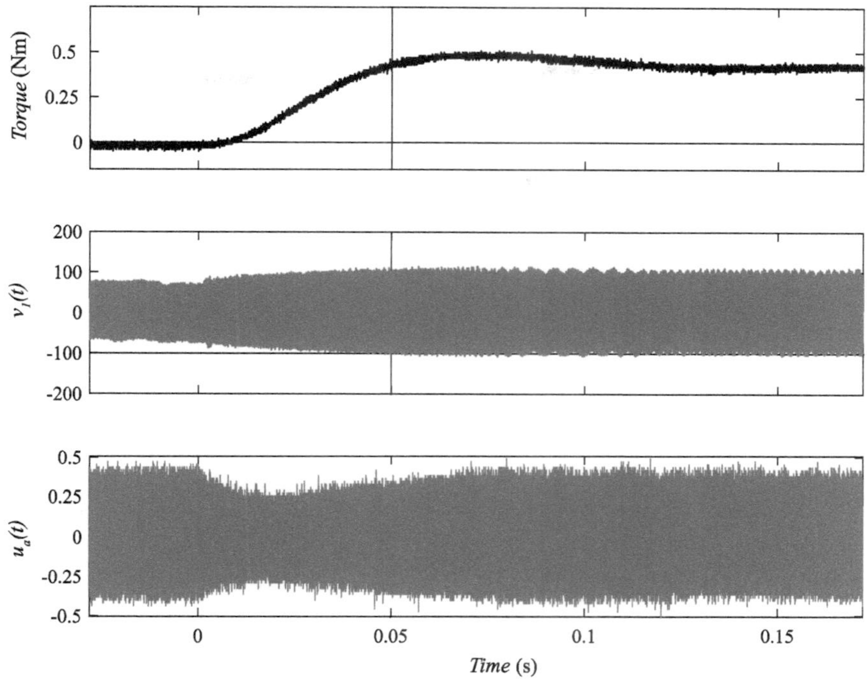

Figure 5.14 Transitory response of the TWUM to a step variation of the load torque *T*; the curves are presented as they are measured with an oscilloscope.

We can check the robustness of the control to the load torque T_r. For this, we apply a braking torque equal to 0.5 Nm in 100 ms on the TWUM's shaft, while the reference U_{dref} remains constant and approximately equal to 0.5 m/s. The results are depicted in Fig. 5.14, in the temporal domain.

This figure shows the behavior of the motor. When increasing the load torque, the vibration velocity amplitude U_{ad} decreases, because voltage V_{1d} has to be shared between the vibration and the torque. This reduction in vibration creates an error which is compensated by the controller in axis *d*; its action is to increase V_{1d} from 76 V, when operating with no load, to 112 V, when the motor is loaded.

Hence, without any torque sensors, VCM automatically adapts its voltage to the amount of torque. The controllers apply lower voltage when no load is applied, and increase it when the motor is loaded. This gives rise to efficiency improvement of the system. This advantage is obtained without more sensors than are available on current TWUMs.

CHAPTER SIX

Practical implementation

Contents

6.1. Introduction 177
6.2. Embedding VCM into a microcontroller 178
 6.2.1 General architecture 178
 6.2.2 The direct digital synthesis 179
 6.2.3 Demodulation of the vibration velocity 182
 6.2.4 Control 184
6.3. Vibration sensor, amplifier and calibration 187
 6.3.1 Measuring the vibration velocity 187
 6.3.2 Amplifier 190
 6.3.3 Calibration 192
6.4. Simplification for $V_q = 0$ and a switching converter 195
 6.4.1 Principle of the voltage supply 195
 6.4.2 Principle of the measurement of U_d and U_q 197
 6.4.3 Implementation in a microcontroller 198

Abstract

In this chapter, we present an example of how we can implement the vector control method into a microcontroller or a DSP. Some source codes are given for creating the rotating reference frame and the controllers presented in Chap. 4. Then, we show how this implementation simplifies when specific hardware is used.

Keywords

vector control method, microcontroller, direct digital synthesis, demodulation, calibration, implementation

6.1. Introduction

The previous chapters have demonstrated how the vector control method (VCM) can be used with ultrasonic transducers in order to control their vibration amplitude. We have seen experimental results obtained with actual devices, used in several applications. Therefore, this method is general, and once the principle of the closed-loop control is known, and the parameter identification understood, the application of the method is quite straightforward, provided that we can measure the vibration and apply

https://doi.org/10.1016/B978-0-12-814186-1.00010-7

the voltage into the rotating reference frame, which is the purpose of this chapter.

In fact, the objective here is to describe how to implement this method in a cost-effective microcontroller. The solution is not unique, other solutions are possible, but we present a way which works for ultrasonic transducers operating at 100 kHz and below. Indeed, we take advantage of microcontrollers of the new generation, which embed high sampling rate analog-to-digital converters (ADCs) up to mega sample per second (MSPS), high resolution pulse width modulator (PWM) in addition with 32 bit capability for the arithmetic unit (AU). The fact that all these peripherals are inside the same device allows a key requirement of the vector control method: the synchronization between the reference of voltage $v(t)$ and the measurement of vibration velocity $u(t)$.

In the following section, we show the general architecture of the programming of a microcontroller embedding the VCM. Then, we will present designs of an amplifier and sensor in order to connect this microcontroller to the transducer. Finally, we show that the architecture can be simplified in specific circumstances, at the expense of performance reduction.

6.2. Embedding VCM into a microcontroller

6.2.1 General architecture

Fig. 4.6 on page 142 in Chap. 4 shows the two transformations used for VCM control: the $dq \rightarrow Im$ actually acts as a modulation. Indeed, the output signal of this block, namely $N\_v(t)$ is a modulated digital signal with carrier pulsation ω. This signal is converted by an ADC to create voltage V_i, which is amplified to create $v(t)$. In turn, the $Im \rightarrow dq$ block converts a sinusoidal signal at the same pulsation ω into two signals which are functions of its amplitude and phase. This operation can be achieved through a demodulation of $V_u(t)$, a voltage which is proportional to $u(t)$, so at pulsation ω. Therefore, the implementation should include:

- a signal modulation,
- a signal demodulation,
- a carrier signal generator at frequency $\omega/2\pi$,
- an analog-to-digital conversion (ADC) to measure $V_u(t)$, with N_{ADC} resolution,
- a digital-to-analog conversion (DAC) to output $V_i(t)$, with N_{DAC} resolution,

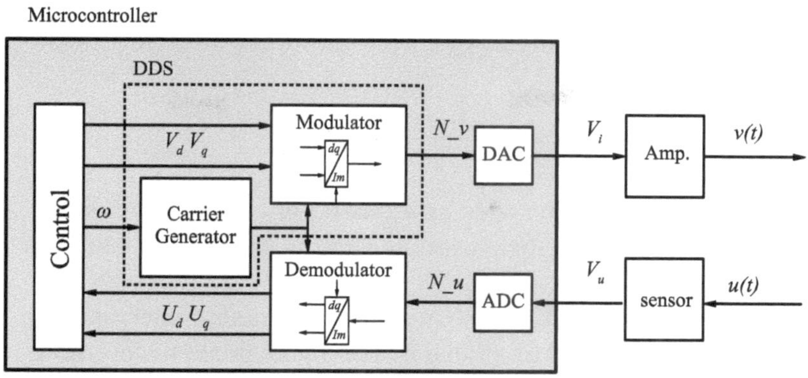

Figure 6.1 VCM is implemented using modulation/demodulation technique; a direct digital synthesizer (DDS) is implementing 2 blocks.

with a representation given in Fig. 6.1.

In this implementation, the signal modulation and generator are synthesized with a direct digital synthesizer (DDS) which is presented in the next section. The other blocks will be presented in the following sections.

6.2.2 The direct digital synthesis

The DDS is made of three blocks:

- a timer which produces a precise clock at a chosen period named T_{DDS},
- an N_{DDS}-bits accumulator called A,
- a table containing the 2^{N_p} values of a cosine, coded on N_t bits; this table is called the *cos-table* C; usually, $N_t = 8$.

The frequency of the carrier signal is programmed through an N_{DDS}-bits number called F.

The algorithm of the DDS is as follows: at each period of the timer, F is added to A. The N_p highest order of A are considered as a register called I, which is used as the index of the *cos-table*. Hence, at each period of the timer, the value of A increases, and so does the index of the *cos-table* I. The period of I is given by the time at which A overflows, leading to Eq. (6.1):

$$\begin{aligned} T &= \frac{2\pi}{\omega} = \frac{2^{N_{DDS}}}{F} \times T_{DDS}, \\ f &= \frac{1}{T} = \frac{F}{2^{N_{DDS}}} \times \frac{1}{T_{DDS}}. \end{aligned} \tag{6.1}$$

The key advantage of the DDS is that the frequency f can be precisely chosen, with a resolution of

$$\Delta f = \frac{1}{2^{N_{DDS}} T_{DDS}}. \tag{6.2}$$

By shifting the index of the table by a quarter of a period, we obtain the value of sine for the same angle with the same table. The index for the sine is named J.

Fig. 6.2 shows a typical DDS operation. At each timer's interruption, the accumulator A is increased by a value F. This figure shows the overflow of F which happens at the 21st event of the timer in this example. Index I is also represented with a bold line, on a different vertical axis scale. Index I has 8 values (3 bits), therefore the *cosine* table has 8 values, too. Due to the value of F in this example, 2 or 3 timer periods are needed to change the value of I. Therefore, the preservation of the form of $V_i(t)$ is not guaranteed from one period of the synthesized signal to another. In particular, with the DDS there is no need for a fixed and integer number of timer periods during a period of the synthesized signal.

By definition of the vector control, Eq. (3.15) on page 104 is recalled here:

$$\begin{aligned} v(t) &= \Im\left(\left(V_d + jV_q\right)e^{j\omega t}\right) \\ &= V_d \sin(\omega t) + V_q \cos(\omega t). \end{aligned} \tag{6.3}$$

In the remaining of this chapter, we use the following convention: $[X]$ is the number X in the memory of the microcontroller. For instance, the DDS generates $\cos(\omega t)$ coded on N_t bits, so we have $[\cos(\omega t)] = 2^{N_t} \times \cos(\omega t) = C(I)$. Moreover, $[\sin(\omega t)]$ is calulated by using an index J which is equal to $I + \frac{2^{N_{DDS}}}{4}$ which corresponds to a phase shift of $\frac{\pi}{2}$.

Eq. (6.3) could be used to calculate N_v at each timer period according to the values $[V_d]$ and $\left[V_q\right]$:

$$N\_v = [V_d] \times C(J) + \left[V_q\right] \times C(I). \tag{6.4}$$

However, Eq. (6.4) is not suitable in a microcontroller, in particular because it can produce a negative integer. Indeed, N_v actually is the input of an ADC, and it has to be a positive integer. Therefore, the calculation of N_v should also include:

- a scaling in order to adapt the range of the output voltage to the full range of the DAC,

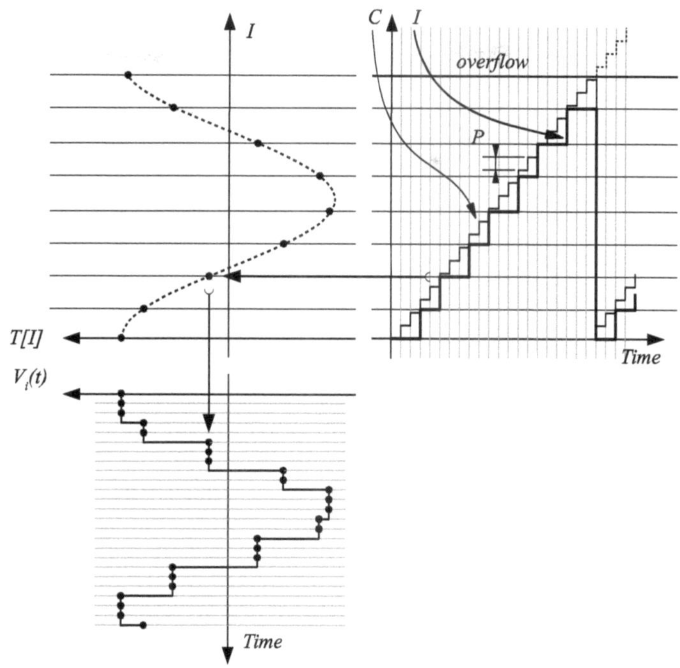

Figure 6.2 Principle of the DDS.

- a biasing in order to convert negative values; generally, the bias is chosen at mid-range of the DAC.

Hence, taking into account these remarks, Eq. (6.4) is revised into:

$$
\begin{aligned}
N\_v &= 2^{N_{DAC}} \times \left(\frac{[V_d] \times C(J) + [V_q] \times C(I)}{G_{DAC}} + \frac{1}{2} \right) \\
&= \frac{2^{N_{DAC}}}{G_{DAC}} \times \left([V_d] \times C(J) + [V_q] \times C(I) + \frac{G_{DAC}}{2} \right)
\end{aligned}
\tag{6.5}
$$

where G_{DAC} is the scaling gain, corresponding to the maximal value of the result of Eq. (6.4).

For the purpose of illustration, we give the code which can be programmed into a microcontroller to generate the output N_v; A is coded on an unsigned long int ($N_{DDS} = 32$), I is on an unsigned char ($N_p = 8$), $[V_d]$, $[V_q]$ are coded on signed long integers (32 bits). In this example, table C is stored in a global table. The DAC has $N_{DAC} = 12$ bit resolution, and we suppose that at full scale, the result of (6.4) is the maximal long integer ($G_{DAC} = 2^{31}$):

```
function long DDS(unit F, long V_d, long V_q){
  private static ulong A=0;
  unsigned char I;
  long N                    % intermediate result

  A=A+F;                    % increasing the DDS accumulator
  I = A >> (24);            % taking the first 8 bits for the index
  J = (A+2^30) >> (24);     % shifting the index to calculate sine

  N = V_d*C[J]+V_q*C[I] + 2^30;   % C stores 2^8 bits of the cosine
                                  % values G_DAC/2 = 2^30

  return N >> (19);  % equivalent to dividing by 2^19=G_DAC/2^N_DAC
}
```

Hence, the DDS can generate the signal defined by V_d and V_q in the rotating reference frame. The angle ωt is given by register A. In the next section, the DDS is used to calculate the values U_d and U_q of the measured vibration velocity.

6.2.3 Demodulation of the vibration velocity

Today's microcontrollers can embed high speed analog-to-digital converters (ADC) which can sample a signal at high frequency, for processing at lower rate. The demodulation of $u(t)$ takes benefit of these technical advances.

Eq. (3.15) gives

$$u(t) = U_d \sin(\omega t) + U_q \cos(\omega t). \tag{6.6}$$

Moreover, the following integral calculations give:

$$\begin{aligned} \frac{2}{T}\int_0^T u(t)\sin(\omega t)\mathrm{d}t &= \frac{2}{T}\int_0^T \big(U_d \sin(\omega t) - U_q \cos(\omega t)\big)\sin(\omega t)\mathrm{d}t \\ &= U_d \end{aligned} \tag{6.7}$$

and

$$\begin{aligned} \frac{2}{T}\int_0^T u(t)\cos(\omega t)\mathrm{d}t &= \frac{2}{T}\int_0^T \big(U_d \cos(\omega t) - U_q \sin(\omega t)\big)\cos(\omega t)\mathrm{d}t \\ &= U_q. \end{aligned} \tag{6.8}$$

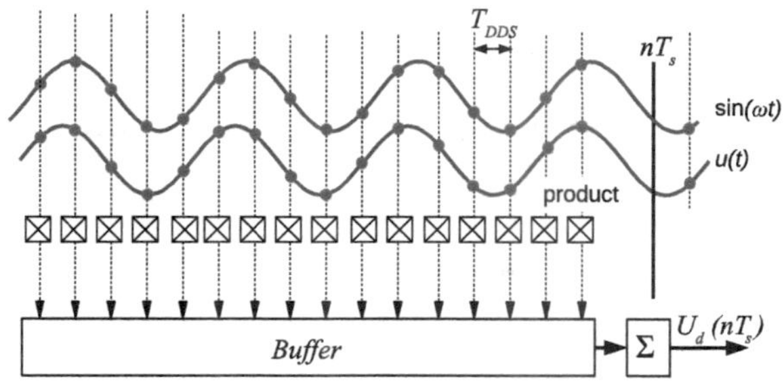

Figure 6.3 Principle of the demodulation of U_d with $N_D = 3$; • correspond to the sampled values of $u(t)$ and $\sin(\omega t)$.

The demodulation of U_d and U_q is then achieved with a discrete integral: at each timer period of the DDS, the two products $u(t) \times \sin(\omega t)$ and $u(t) \times \cos(\omega t)$ are calculated, and the results are stored in a buffer memory. When the memory has stored enough product values – we call N_D the number of values – their summation will directly give U_d and U_q, provided that N_D corresponds to a multiple of period T. This value can be sampled every $T_s > N_D \times T$ by the vector control method algorithm in order to control the vibration velocity. This is explained by Fig. 6.3.

A key advantage of this demodulation technique is that it rejects the higher modes which can be measured by the vibration velocity sensor. Indeed, for sake of clarity and without lack of generality, we consider a vibration velocity which contains two pulsations ω and ω_u such that

$$u(t) = U_d \sin(\omega t) + U_u \sin(\omega_u t). \tag{6.9}$$

Then, the integral in (6.7) calculated over N_D period gives

$$\begin{aligned} &\frac{2}{N_D T}\int_0^{N_D T} u(t)\sin(\omega t)\mathrm{d}t \\ &= \frac{2}{N_D T}\int_0^{N_D T} \left(U_d \sin(\omega t) + U_p \sin(\omega_u t)\right)\sin(\omega t)\mathrm{d}t \\ &= U_d + U_p \frac{1}{\pi N_D}\sin(2\pi N_D \frac{\omega_u}{\omega})\frac{\omega_u \omega}{\omega_u^2 - \omega^2}. \end{aligned} \tag{6.10}$$

Eq. (6.10) shows that when demodulating a signal containing the contribution of an unwanted mode at ω_u, an error is made. We calculate the ratio R of this error as

$$R = \frac{1}{\pi N_D}\sin(2\pi N_D \frac{\omega_u}{\omega})\frac{\omega_u \omega}{\omega_u^2 - \omega^2}. \tag{6.11}$$

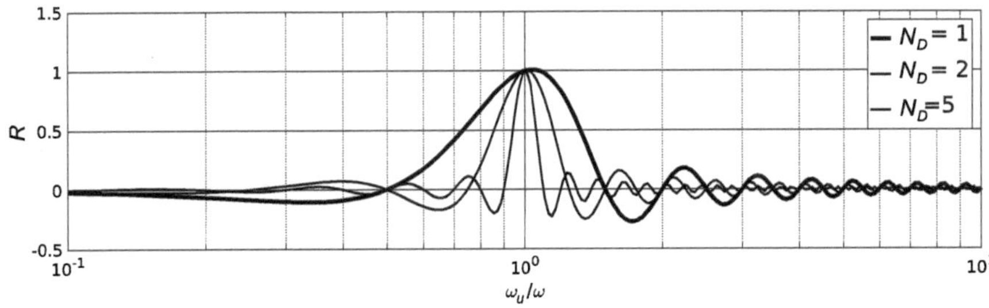

Figure 6.4 Error ratio R as a function of ω_U/ω for $N_D = 1, 2$ and 5.

If $R = 0$ then the mode has no effect, and (6.7) will give U_d. If $R = 1$ then the unwanted mode can contribute to the estimate of U_d with the same gain as the mode at ω. We draw, in Fig. 6.4, R as a function of $\frac{\omega_u}{\omega}$.

This figure shows that other modes may create an error to the estimate of U_d because R does not equal 0 everywhere. However, for $N_D = 1$, the error is limited ($-0.1 < R < 0.1$) for $\omega_u/\omega > 3.3$ or $\omega_u/\omega < 0.3$. If more periods can be taken into account for the calculation of U_d the value is more robust to the other modes. Indeed, with $N_D = 2$, the error is limited for $\omega_u/\omega > 1.89$ or $\omega_u/\omega < 0.569$ and for $N_D = 5$, the error is limited for $\omega_u/\omega > 1.35$ or $\omega_u/\omega < 0.75$.

6.2.4 Control

In this section, we implement VCM into a microcontroller. For the sake of conciseness, we will present the control of axis d only, without taking into account the coupling between axes q and d. The figure of the closed-loop control in axis d is recalled in Fig. 6.5.

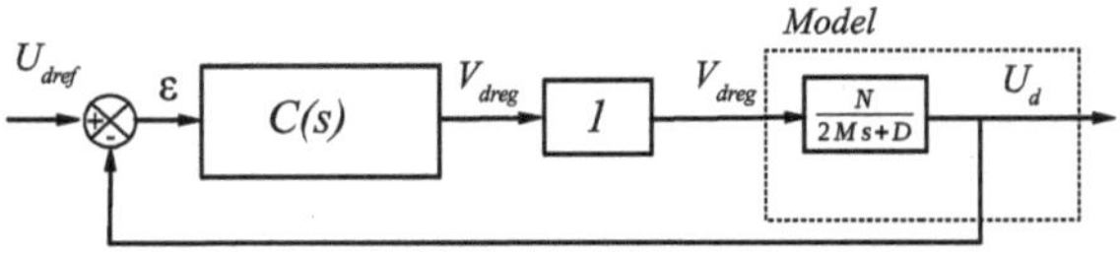

Figure 6.5 Control of axis d; it is assumed that $V_{dreg} = V_d$, meaning that the calculated value of V_{dreg} can be imposed on the transducer.

This control scheme cannot be directly implemented. Indeed, microcontrollers and DSPs use numbers to represent the variables. These numbers can be integers, or float. For a cost-effective implementation, it is recommended to use integers only, because float numbers require more resources. We control a system which is connected to the real world, and

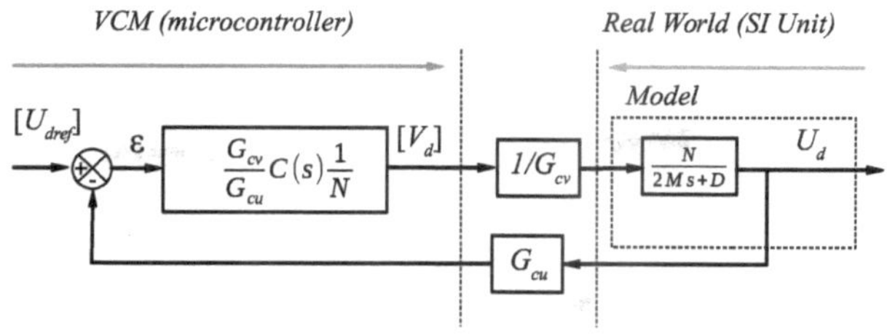

Figure 6.6 Control of axis *d*, with conversion factor.

the variables have a dimension. For instance, V_d and V_q are in volt, U_d and U_q are in m/s. The dimension is not coded in a DSP, and, without any specific precaution, the variables are expressed in the International System of Units (SI). But integers have a fixed resolution (it is 1), therefore, for the typical values of voltage and vibration velocity which are measured experimentally, this resolution is too big. For instance, a transducer vibrating at $U_d = 0.1$ m/s becomes $U_d = 0$ with integers!

Therefore, it is necessary to change the dimension of the variables, and to *convert* them using a conversion factor G_C. For example, we can decide to express the vibration velocity in hundreds of micrometers per second; using this unit, a vibration speed $U_d = 0.1$ m/s is coded as $U_d = 1000$. The conversion factor is $G_{cu} = 10\,000$. This way, the resolution of U_d is 100 μm/s and not 1 m/s. Other conversion factors are possible, it depends on the application, the minimal resolution it requires, and the number of bits of the microcontroller. The voltage too is expressed with a higher resolution (for example, in microvolts and $G_{cv} = 1\,000\,000$).

Therefore, we will have:

$$[V_d] = G_{cv} V_d \quad \text{and} \left[V_q\right] = G_{cv} V_q \quad \text{and} [v] = G_{cv} v \tag{6.12}$$

and

$$[U_d] = G_{cu} U_d \quad \text{and} \left[U_q\right] = G_{cv} U_q \quad \text{and} [u] = G_{cv} u. \tag{6.13}$$

When adding these gains, we change the open-loop transfer function. It is then necessary to take into account this fact in the calculation of the corrector. For this, the parameters of $C(p)$ are multiplied by $\frac{G_{cv}}{G_{cu}}$ as explained in Fig. 6.6.

One will verify that the open-loop transfer function is unchanged compared with that of Fig. 6.5, so $C(s)$ is unchanged, too.

Chapter 4 proposed to us *a PI* controller for $C(s)$ and detailed the calculation of its parameters K_p and K_I. But how can we implement $C(s)$ in a microcontroller? The response is a 2-step process.

First, we discretize the controller using the Tustin transform. If we consider a sampling period T_s then the controller is discretized by replacing the Laplace variable s by

$$s \leftarrow \frac{2}{T_s}\frac{1-z^{-1}}{1+z^{-1}}, \tag{6.14}$$

which gives for the complete controller of axis d:

$$\begin{aligned} \frac{[V_d(z)]}{\epsilon(z)} &= \frac{G_{cv}}{G_{cu}}N\left(K_p + \frac{K_I T_s}{2}\frac{1+z^{-1}}{1-z^{-1}}\right) \\ &= \frac{G_{cv}}{G_{cu}}N\frac{\left(K_p+\frac{K_I T_s}{2}\right) - z^{-1}\left(K_p - \frac{K_I T_s}{2}\right)}{1-z^{-1}}, \end{aligned} \tag{6.15}$$

leading to

$$[V_d(z)](1-z^{-1}) = \frac{G_{cv}}{G_{cu}}N\left(\left(K_p + \frac{K_I T_s}{2}\right) + z^{-1}\left(K_p - \frac{K_I T_s}{2}\right)\right)\epsilon(z). \tag{6.16}$$

Using the shift theorem, we can express the new value of $[V_d]$, labeled $[V_d(k)]$, according to the previous value $V_d(k-1)$ and the actual and previous values, $\epsilon(k)$ and $\epsilon(k-1)$,

$$[V_d(k)] = [V_d(k-1)] + A\epsilon(k) - B\epsilon(k-1) \tag{6.17}$$

with $A = \frac{G_{cv}}{G_{cu}}N\left(K_p + \frac{K_I T_s}{2}\right)$ and $B = \frac{G_{cv}}{G_{cu}}N\left(K_p - \frac{K_I T_s}{2}\right)$.

The following code is an example of how the controller can be implemented: A and B are constants, while $\epsilon(k-1)$ and $V_d(k-1)$ are stored in a static memory location.

```
function long sample_axisd(int Udref, int Ud){
      int eps;                          // eps;
      private static int old_eps = 0;   // eps(k-1)
      private static long old_Vq=0;     // Vd(k-1)
      long Vd=0;

      \\definition of the corrector's gain
```

```
        #define A (long)(Gcv*Gcu*N*(Kp+Ki*Ts/2))
        #define B (long)(Gcv*Gcu*N*(Kp-Ki*Ts/2))

        eps = (Udref - Ud);
        Vd = old_Vd + A*eps - B*old_eps;        // PI controller

        old_Vd = Vd; old_eps = eps;
return Vd
}
```

6.3. Vibration sensor, amplifier and calibration

6.3.1 Measuring the vibration velocity

Using laser interferometers is an expensive way to measure the vibration velocity of a transducer, and they are reserved for applications which require precision and resolution. In our industrial processes, we need cheap and robust vibration sensors instead. A good solution is often to place a small piece of piezoelectric material at the right place to obtain, by the *direct* piezoelectric effect, a voltage proportional to the displacement.

Indeed, when the material undergoes a deformation, a current i_m is produced as already presented in Fig. 1.4 on page 5. We have demonstrated that $i_m = Nu$ where i_m is called the *motional* current, N is the force factor, and u is the vibration velocity. If the electrodes of the piezoelectric element are left open, then i_m flows into the internal capacitance C, and a voltage v_m is produced across the electrodes:

$$v_m = \int CNu\mathrm{d}t = NC \int \frac{\mathrm{d}w}{\mathrm{d}t}\mathrm{d}t = NCw(t). \tag{6.18}$$

Hence, when the terminals of the piezoelectric elements are left open, the voltage v_m is proportional to the displacement w.

From that measurement, the vibration velocity can be deduced by numerical derivation of v_p. However, this solution is not practical because numerical differentiation amplifies noise and requires execution time on the processor. Therefore, it is necessary to measure the vibration velocity directly. For this, we connect a resistor R_m on the element's terminals. If the value of R_m is small enough, then the impedance of C at the vibration pulsation can be higher than R_m ($R_m \ll 1/C\omega$). Under this condition, current

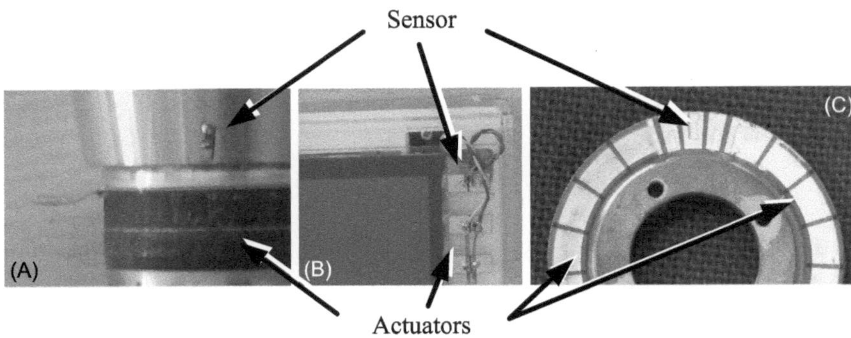

Figure 6.7 Implementations of a vibration velocity sensor: (A) on a Langevin transducer, (B) in a haptic surface, and (C) on Traveling Wave Ultrasonic Motor.

i_m flows into resistor R_m, and we get

$$v_m = R_m i_m = R_m N u. \tag{6.19}$$

In practice, the value of R_m is chosen high enough to obtain a measurable voltage on the sensor's terminal. Moreover, the location of the sensor has to be adequately chosen. Indeed, it is convenient to put the piece of piezoelectric material in place where the force factor N is the highest. This location corresponds to where the piezoelectric material should be placed for the actuator mode, because also in actuator we want to maximize N. Hence, the best location of the vibration velocity sensor coincides with that of the actuator.

Fig. 6.7 shows several implementations of a vibration velocity sensor. On the left-hand side of the figure, we have a Langevin transducer. The actuators are aside the vibration node, and adding a sensor is not possible without dismantling; gluing a small piezoelectric patch on the counter mass is not the best place, but at least allows the measurement of the vibration velocity. The example in the middle presents a velocity sensor for a haptic surface and is the same plate as that of the actuators'. The last example is taken from the stator of a Travelling Wave Ultrasonic Motor. A ring of piezoelectric material is glued on it, and to create the velocity sensor, the electrode reserve a small surface to be used as for the sensor.

The issue in this method is to share the current between the motional current and the current flowing into the internal capacitance C. For that purpose, a wheaston bridge can be used, as shown Fig. 6.8. If we denote by v and i the voltage and current of the piezoelectric transducer, the circuit

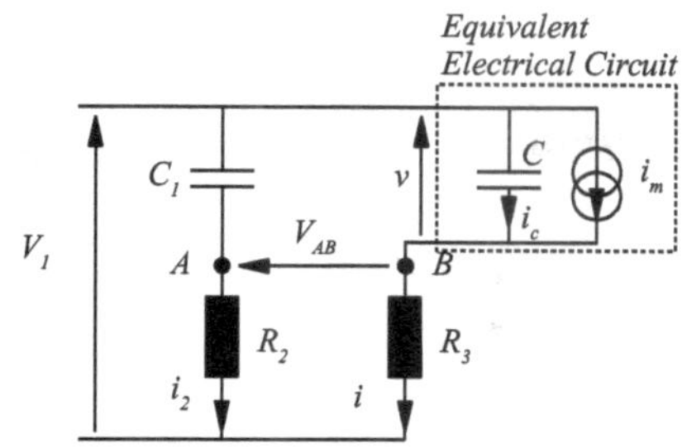

Figure 6.8 Implementations of a Wheaston bridge to measure the motional current i_m.

gives

$$V_B = R_3 i = R_3(i_m + jC\omega v). \tag{6.20}$$

In this circuit, the power is supplied by a voltage source V_1. Hence, we have

$$v = V_1 - V_B. \tag{6.21}$$

Substituting (6.20) into (6.21) leads to

$$V_B = \frac{R_3}{1 + jR_3 c\omega} i_m + \frac{jR_3 C\omega}{1 + jR_3 C\omega} V_1. \tag{6.22}$$

The circuit $R_2 C_1$ is a voltage divider, so we have

$$V_A = \frac{jR_2 C_1 \omega}{1 + jR_2 C_1 \omega} V_1. \tag{6.23}$$

At the end, the voltage $V_{AB} = V_A - V_B$ is equal to

$$V_{AB} = \left(\frac{jR_3 C\omega}{1 + jR_3 C\omega} - \frac{jR_2 C_1 \omega}{1 + jR_2 C_1 \omega} \right) V_1 - \frac{R_3}{1 + jR_3 C\omega} i_m. \tag{6.24}$$

If the values of C_1, R_2 and R_3 are adequately chosen, such that $R_2 C_1 = R_3 C$, Eq. (6.24) gives

$$V_{AB} = -\frac{R_3}{1 + jR_3 C\omega} i_m. \tag{6.25}$$

Then, if $R_3 C\omega \ll 1$, Eq. (6.25) simplifies to

$$V_{AB} = -R_3 i_m = -R_3 N u. \tag{6.26}$$

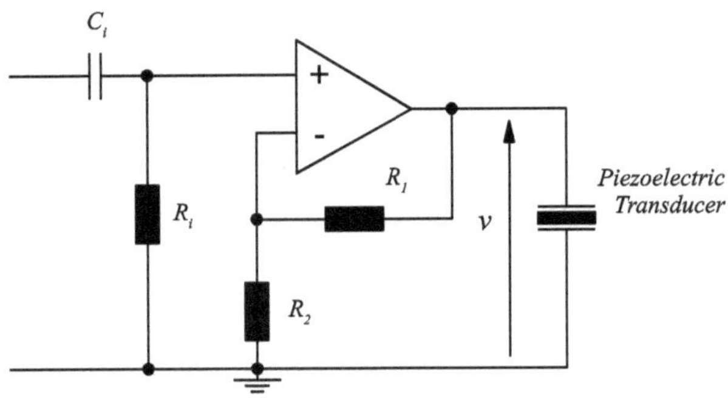

Figure 6.9 A voltage amplifier.

6.3.2 Amplifier

When the number N_v is calculated by the DDS, this value is converted by an analog voltage, thanks to the *digital-to-analog converter* to create V_i. This voltage is then amplified, because most of the time the controller cannot supply enough current to the piezoelectric load. Since the vector control can track the resonance frequency of the transducer, we can use a voltage amplifier to supply power to the device. Indeed, at resonance, less voltage is needed compared with operation outside resonance, and linear amplifiers become then attractive. The classical realization of a voltage amplifier based on an operational amplifier is given in Fig. 6.9.

Additional components are added for several purposes. First, the output of a DAC is biased to evolve between 0 V and V_{DD}. It is then mandatory to remove this bias before amplification. This is achieved by R_iC_i which acts as a high-pass filter, and we will choose $R_iC_i\omega \ll 1$ such that in principle it has no effect at the working pulsation.

The two resistors, R_1 and R_2, fix the gain of the amplifier, which is given by

$$v = G_A \times V_i = \left(1 + \frac{R_1}{R_2}\right) V_i. \tag{6.27}$$

Gains up to $G_A = 20$ can be achieved with this circuit. However, the efficiency of voltage amplifiers is low, in particular because, by their working principle, they create a voltage drop to control the output voltage. It is therefore important to reduce the current to limit the power losses inside the amplifier. To do so, an inductor can be placed in parallel to the amplifier's output (Svilainis and Motiejunas, 2006). The value of the inductance L_m is chosen so that its impedance corresponds to that of the capacitance C

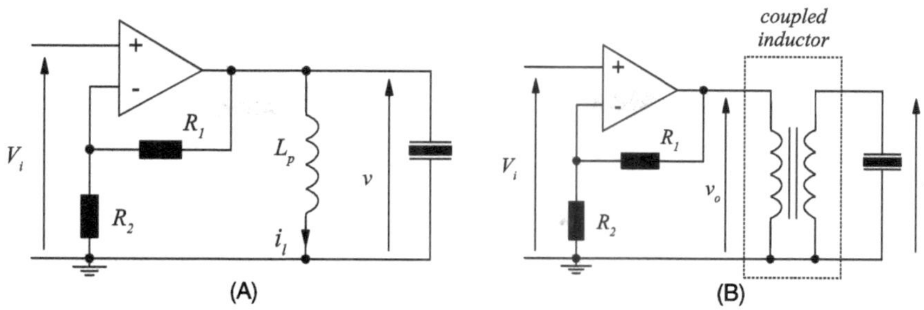

Figure 6.10 Voltage amplifier with (A) output parallel inductor and (B) a *coupled inductor*; compared with Fig. 6.9, some parts have been removed for clarity.

at the working frequency, leading to $L_pC\omega^2 = 1$. In that condition, current i_l in the inductor perfectly compensates the current in the internal capacitance i_c at the working frequency; as a consequence, the output current of the voltage amplifier is reduced to the motional current i_m. This solution is depicted in Fig. 6.10 (A).

To further increase the voltage on the transducer, inductor L_p can be coupled to another inductor to form a *coupled inductor*. A *coupled inductor* looks like a transformer but with a low magnetizing inductor. Hence, voltage v equals k_Tv_o where k_T is the transformer ratio, v_o is the output voltage of the amplifier, while the condition to fulfill for the magnetizing inductor is given by $L_pCk_T^2\omega^2 = 1$ to take into account that the inductor is at the primary, while the transducer is at the secondary of the coupled inductor. When this condition is obtained, the output current of the amplifier is equal to k_Ti_m.

A voltage amplifier actually achieves a closed-loop control, and therefore, it can be unstable (Apex Microtechnology, 2013). Two main reasons are raised to explain the instabilities. First, the internal impedance of the amplifier reduces the phase margin, especially on large capacitive loads; the effect is prominent with low gain ratio. Second, ground loops can re-inject a fraction of the output with a positive feedback. Since ultrasonic transducers operate mostly at frequencies above 20 kHz, it is necessary to speed-up the amplifier. Due to the so-called stability-performance of a closed-loop control, increasing the bandwidth of the amplifier also makes it more prone to instabilities. It is then necessary to stabilize the amplifier, and a good way is to place an R_sL_s circuit in series with the output. This circuit acts as a low-pass filter, and the values of R_s and L_s have to be tuned in order to place its corner frequency above the working frequency. For example, Fig. 6.11 (A) shows the frequency response of the amplifier without the

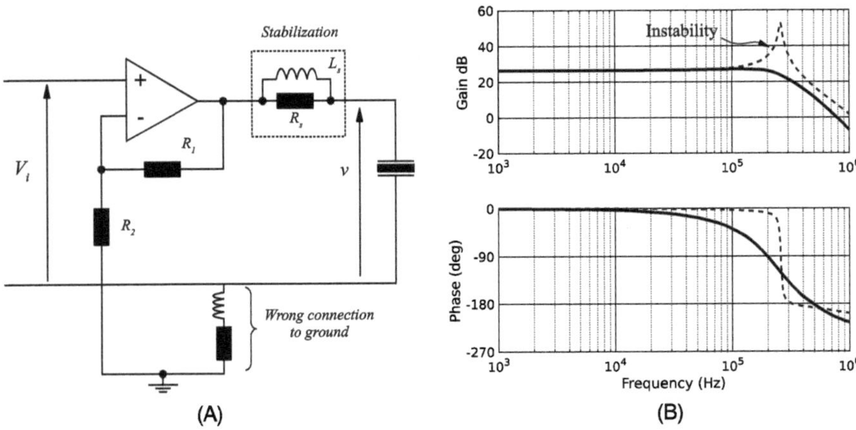

Figure 6.11 Stabilization of a voltage amplifier on capacitive load: (A) shows the circuit and (B) illustrates the frequency response of the amplifier with (–) and without (- -) the stabilization $R_S L_S$ circuit, when a grounded loop occurs. Simulation were obtained with $4\ \Omega$ and $0.5\ \mu$H wrong grounding, an $R_S = 47\ \Omega$, $L_S = 250\ \mu$H stabilization circuit, an LM12 (National Semiconductor) voltage amplifier mounted with a gain of $G_A = 20$, and a transducer with $C = 9$ nF capacitance.

stabilization circuit when a wrong grounding occurs. We see that the circuit has a resonance at 100 kHz, producing ringing of the output voltage to this frequency. When the stabilization circuit is inserted, the resonance is damped, thus stabilization is effective.

This section has shown the basic guidelines of the design of a power amplifier for an ultrasonic transducer. In fact, the tracking of the resonant frequency helps to decrease the voltage demand, and a linear amplifier is an attractive solution. However, stabilization and efficiency improvement add parts, which introduce a phase lag between the amplifier's input and the actual voltage applied to the piezoelectric transducer. In general, this lag cannot be neglected in our particular case, which implements the vector control, because it means that the rotating reference frame in the microcontroller is not in phase with the rotating reference frame of the transducer.

The next section deals with this issue by showing how a calibration can put in phase these two rotating reference frames when a phase lag exists.

6.3.3 Calibration

The calibration process' objective is to make the value of a variable of the system in the microcontroller equal to the value of the same variable measured in the physical world. For instance, the vibration velocity is measured by a sensor, which converts the vibration into a voltage with a specific gain. This voltage is converted to a digital signal, which is demodulated. The system is calibrated, for instance, if $\sqrt{[U_d]^2+[U_q]^2}$ is equal to the amplitude of the vibration velocity measured with a laser interferometer. For this, a calibration gain G_{calu} multiplies the output of the demodulator in order to achieve this objective.

Calibration of $v(t)$ is not so simple. Indeed, the digital signal N_v is converted into a voltage V_i by the DAC, and we have

$$V_i = GN\_v \tag{6.28}$$

where $G=\frac{Vcc}{2^{N_{DAC}}}$, with Vcc being the supply voltage of the amplifier. Then this voltage V_i is stepped-up by the amplifier, which introduces a gain G_A and a phase lag φ_A at pulsation ω. As a consequence, taking into account all these gains, we have

$$\begin{aligned} V_i &= \frac{Vcc}{2}+\frac{Vcc}{G_{DAC}}\left([V_d]\,C(J)+[V_q]\,C(I)\right) \\ &= \frac{Vcc}{2}+\frac{VccG_{cu}}{G_{DAC}}2^{N_t}\left(V_d\sin(\omega t)+V_q\cos(\omega t)\right), \end{aligned} \tag{6.29}$$

leading to

$$\begin{aligned} v(t) &= \Im\left(\frac{VccG_{cu}}{G_{DAC}}2^{N_t}\left(V_d+jV_q\right)e^{j\omega t}G_Ae^{-j\varphi_A}\right) \\ &= \frac{G_A\times Vcc\times G_{cu}}{G_{DAC}}2^{N_t}\Im\left(\left(V_d+jV_q\right)e^{-j\varphi_A}e^{j\omega t}\right). \end{aligned} \tag{6.30}$$

We remark that Eqs. (6.30) and (6.3) differ. Indeed, the voltage across the transducer's terminals is multiplied by a gain $G_A\frac{VccG_{cu}}{G_{DAC}}2^{N_t}$, and has a phase lag φ_A. The effect of the gain is to change the open-loop transfer function, which could result in bad closed-loop performance. The effect of the phase lag is more difficult to analyze.

In theory, the phase lag is not constant. In practice, since the transducer operates in a limited frequency range around the resonance frequency, we assume that φ_A is a constant. Eq. (6.30) shows that we can replace $\left(V_d+jV_q\right)$ by $\left(V_d+jV_q\right)e^{-j\varphi_A}$ in Eq. (6.3). Therefore, the phase lag φ_A is equivalent to a rotation in the (d,q) plane with an angle φ_A of the voltage vector. Its effect is to change the position of the vector $\underline{V}$ in the rotating reference frame. For example, consider a transducer for which frequency tracking is implemented. The VCM operates assuming $V_q=0$, so vector $\underline{V}$ is aligned

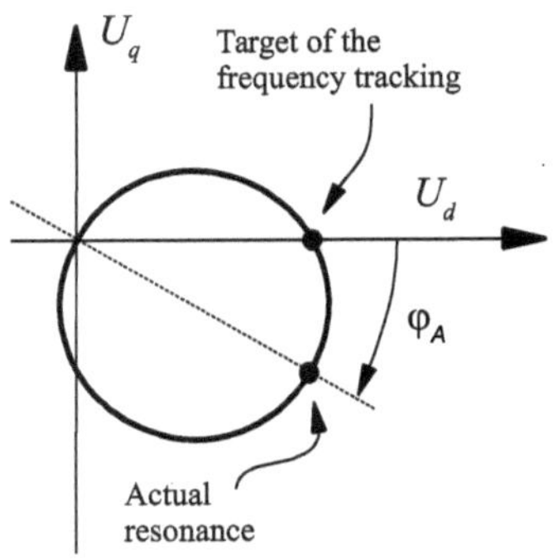

Figure 6.12 Effect of a phase lag φ_A on the extremity of the vector $\underline{U}$ for $V_d = 0$.

with axis d: the vibration velocity amplitude is controlled with V_d, while the pulsation is tuned in order to have $U_q = 0$. But due to the phase lag, actually the position of vector $\underline{V}$ in the rotating reference frame is given by

$$\underline{V} = V_d \cos(\varphi_A) + jV_d \sin(\varphi_A). \tag{6.31}$$

So, $\underline{V}$ is not aligned with axis d, and its component on axis q creates a vibration velocity on axis q: the frequency tracking algorithm will not operate at resonance. This effect is illustrated in Fig. 6.12. We show how the affix of $\underline{U}$ evolves when ω is varied in the vicinity of the resonance pulsation of a transducer with $V_d = 0$ and taking into account a phase lag. Compared with Fig. 3.9 on page 109, it is rotated by φ_A, and clearly, the operating point $U_q = 0$ does not correspond to the maximal vibration velocity (U is maximal). It is therefore mandatory to cancel out the effect of this phase lag.

To calibrate the system, we rotate the calculated vector $\underline{V}$ by an angle φ_A, and we multiply it by a new gain G_{calv}. For this, we introduce V'_d and V'_q such that

$$V'd + jV'_q = G_{calv}(V_d + jV_q)e^{j\varphi_A}. \tag{6.32}$$

If we calculate N_v from $\left[V'_d\right]$ and $\left[V'_q\right]$ and not $[V_d]$ and $\left[V_q\right]$, we will have

$$V_i = \frac{Vcc}{2} + \frac{VccG_{cu}}{G_{DAC}} 2^{N_t} G_{calv} \Im\left(\left(V'_d + jV'_q\right) e^{j\varphi_A} e^{j\omega t}\right), \tag{6.33}$$

leading to

$$v(t) = \frac{VccG_{cu}}{G_{DAC}} 2^{N_t} G_{calv} \Im \left(\left(V_d' + jV_q' \right) e^{j\omega t} \right). \tag{6.34}$$

If the effect of the gain can be compensated, then Eq. (6.34) should be equal to (6.3), which is achieved if

$$G_{calv} = \frac{G_{DAC}}{G_A \times Vcc \times G_{cu} 2^{N_t}}. \tag{6.35}$$

Eq. (6.30) then gives:

$$\begin{aligned} V_d' &= G_{calv} \cos(\varphi_A) V_d - G_{calv} \sin(\varphi_A) V_q = G_1 V_d - G_2 V_q, \\ V_q' &= G_{calv} \sin(\varphi_A) V_q + G_{calv} \cos(\varphi_A) V_q = G_2 V_d + G_1 V_q. \end{aligned} \tag{6.36}$$

The following code is an example of the implementation of the calibration into a DSP. To use integers only, the values of the gains G_1 and G_2 are multiplied by 2^{16} first:

```
#define G1 = (long)(Gcalv * 2^(16)*cos(phiA)) // these 2 lines are
#define G2 = (long)(Gcalv * 2^(16)*sin(phiA)) // interpreted by the compiler
                                              // not executed by the DSP

Vpd = (G1 * Vd - G2 * Vq) >> 16 ; // divide by 2^16 because G1 and G2
Vpq = (G2 * Vd + G1 * Vq) >> 16 ; // are multiplied by 2^16
```

6.4. Simplification for $V_q = 0$ and a switching converter

6.4.1 Principle of the voltage supply

Switching converters have better efficiency than linear amplifiers, and in spite of their high distortion (the voltage is not sinusoidal), they can, however, be utilized to supply power to a resonating load, like a transducer. The VCM is also adapted to this type of converters, as reported in (Maas et al., 1997). In this section, we will use a 2-leg switching converter to supply power on a transducer. The circuit's scheme is detailed in Fig. 6.13. Two inductors with the same value (both denoted L_s) are inserted in order to limit the inrush current when the switches' state changes.

In principle, the two switches, K_i and K_i', $i \in \{1, 2\}$, have complementary states to avoid the shortcut of the voltage source E and the disruption of current i_1 in inductors L_s. When K_1 is closed, one has $V_A = E$, and when K_2 is closed, $V_B = E$. If we assign a Boolean variable to describe the state

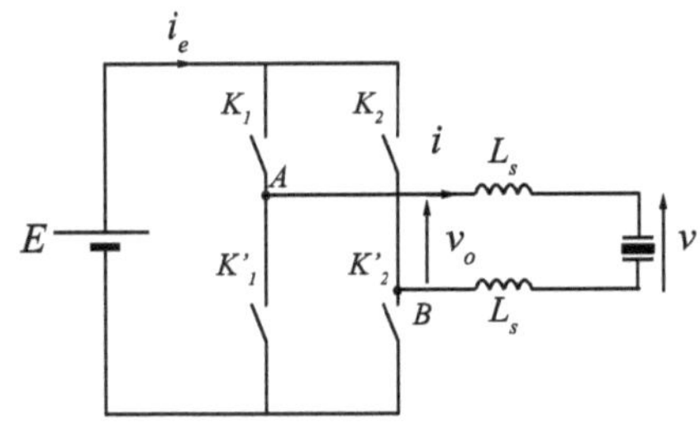

Figure 6.13 The switching amplifier detailed in this chapter.

of the switches, *i.e.*, $K_i = 1$ means it is closed, then we can write

$$V_A = K_1 E \quad \text{and} \quad V_B = K_2 E, \tag{6.37}$$

leading to

$$V_{AB} = V_A - V_B = (K_1 - K_2)E. \tag{6.38}$$

The control of voltage v is achieved by switching on and off the switches periodically, at the same and constant frequency, during the same time $t_\phi = \phi \times T$, as explained in Fig. 6.14.

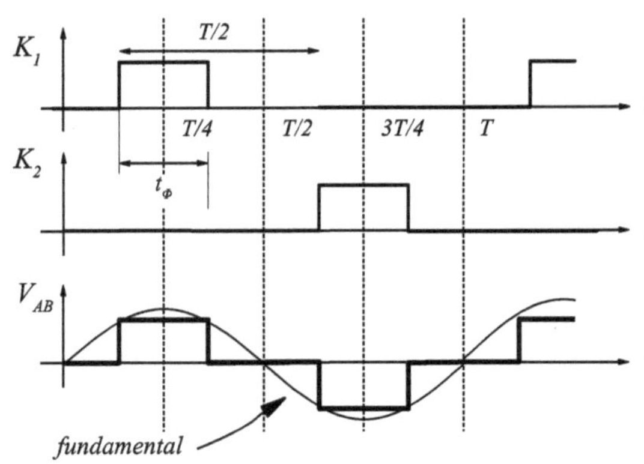

Figure 6.14 Evolution of the switches' states as a function of time, and the resulting voltage V_{AB}; the conduction time t_ϕ is intentionally set symmetrical to $T/4$ and $3T/4$.

Fig. 6.15 presents a simulation of the circuit. As it can be seen, the waveform of $v(t)$ is very similar to that of V_{AB}. The two inductors L_s limit the inrush current to the transducer. Without L_s, the current would have much higher spikes at the commutation of K_1 or K_2, which could dramat-

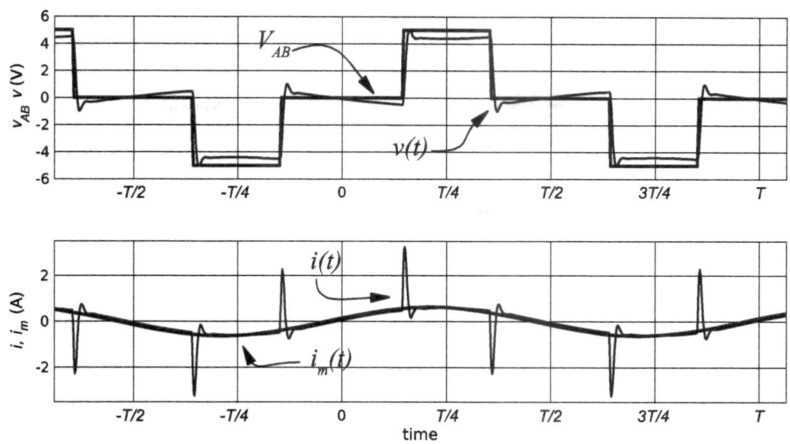

Figure 6.15 Evolution of the V_{AB} (–), $v(t)$ (–), $i_m(t)$ (–), and $i(t)$ (–) for a typical application; $T = 28.5$ μs, $\phi = \frac{1}{4}$, $C = 128$ nF, and $L_s = 0.5$ μH.

ically damage the transducer and/or the converter. However, L_s is chosen sufficiently small to not change the waveform too much.

Although $v(t)$ is not purely sinusoidal, the motional current of the transducer appears to be sinusoidal. In fact, the transducer has a resonant behavior; therefore, it filters out the harmonics of $v(t)$, provided that a vibration mode of the structure is not excited by them. Therefore, the converter is used to control the fundamental of $v(t)$ and not its actual value. For the sake of simplification, we make the assumption that the fundamental of $v(t)$, labeled $v_f(t)$, equals that of V_{AB}, leading to

$$v_f(t) = E \times \sin(\pi\phi) \sin\left(2\pi \frac{t}{T}\right). \tag{6.39}$$

Hence, the amplitude of $v(t)$ can be controlled by adjusting ϕ, while its pulsation can be controlled with T, the period of the switching signals of the switches. However, V_d and V_q cannot be controlled separately with this technique. Therefore, we lay down $V_q = 0$, and the rotating reference frame is aligned on $\underline{v}$.

6.4.2 Principle of the measurement of U_d and U_q

The demodulation has many advantages, but it requires the signal $\cos(\omega t)$ and $\sin(\omega t)$ to calculate U_d and U_q from a large number of sampled values of $u(t)$. When these signals are not available, for instance, to operate the VCM

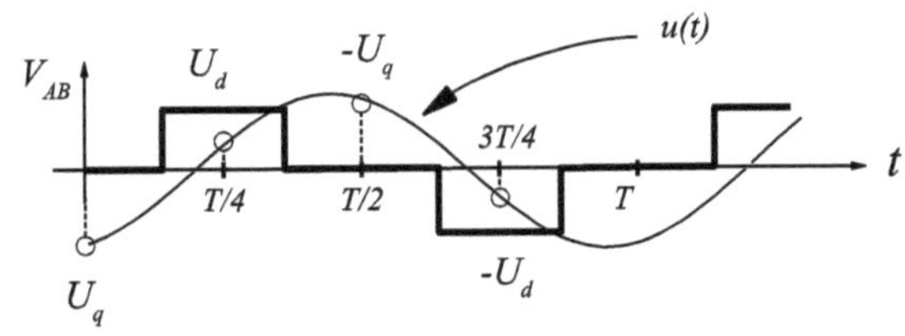

Figure 6.16 Sampling of $u(t)$ in order to measure U_d and U_q.

control in a very cheap microcontroller, we have to use the *synchronized sampling* instead, which is presented hereafter.

Considering Eq. (6.6), we remark that $u(t) = U_d$ when $\cos(\omega t) = 0$, *i.e.*, for $t = T/4$. Hence, sampling u precisely at $t = T/4 + nT$, $n \in \mathbb{N}$ gives the value of U_d. In the same way, sampling u at $t = 0 + nT$, $n \in \mathbb{N}$ gives U_q. The principle of the measurement of U_d and U_q consists then in sampling $u(t)$ every $T/4$, and calculating (Ben Messaoud, 2016)

$$
\begin{aligned}
U_d &= \frac{u(\frac{T}{4}) - u(3\frac{T}{4})}{2}, \\
U_q &= \frac{u(0) - u(\frac{T}{2})}{2}.
\end{aligned}
\tag{6.40}
$$

Since we consider that the rotating reference frame is aligned on $\underline{v}$, the aforementioned sampling times correspond to those presented on the x-axis in Fig. 6.14. Therefore, the operation of an ADC of a microcontroller should be synchronized on the switching of K_1 and K_2 as depicted in Fig. 6.16.

The synchronization of the ADC, together with the generation of the signals K_1 and K_2, is achieved thanks to the operation of a timer, and this is presented in the next section.

6.4.3 Implementation in a microcontroller

A timer offers several features: it generates Boolean signals with controlled period and pulse width; it can also produce hardware interruption based on events, on which action can be triggered. Basically, a timer is a counter with many comparators, as depicted by an accumulator N and a register PR. The timer has several modes: UP means that the counter's register, labeled N, counts from 0 to PR, DOWN; N is decreased from PR to 0. For this implementation, we will use the *UP–DOWN* mode: N is first incremented every clock cycle T_H. Its value is compared to PR, and when

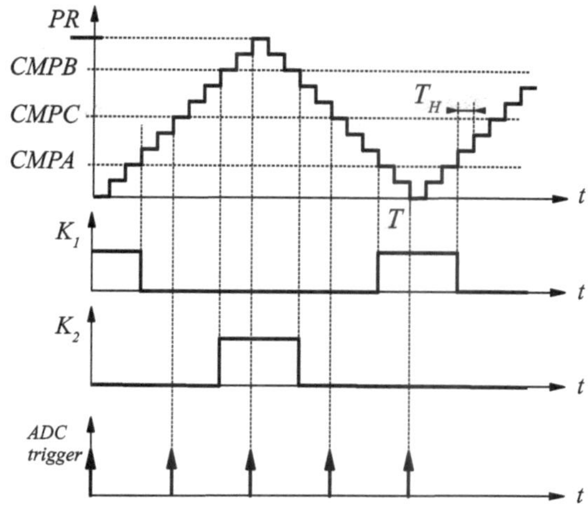

Figure 6.17 Operation of a timer; in this example, $PR = 10$, $CMPA = 2$ and $CMPB = 8$.

$N = PR$, the counting direction is reversed, and N starts to be decremented until $N = 0$, when N is incremented. Therefore, the value of PR defines the period of the timer, and we have

$$T = (2 \times PR + 1) \times T_H. \tag{6.41}$$

Up to 4 comparators can be associated with this timer, named *CMPA*, *CMPB*, *CMPC*, and *CMPD*. The output of these comparators can be associated with outputs and/or interruptions. *CMPA* is associated with the output K_1 such that $K_1 = 1$ if $N < CMPA$. In the same way, *CMPB* is associated with K_2 such that $K_2 = 1$ if $N \geqslant CMPB$. To operate according to the timing of Fig. 6.14, we set *CMPA* and *CMPB* complementary, such that $CMPB = PR - CMPA$. This is illustrated in Fig. 6.17.

Therefore, we have the following relationship between *CMPA*, *PR*, t_ϕ and T_H:

$$t_\phi = (2 \times CMPA + 1) \times T_H, \tag{6.42}$$

leading to

$$\phi = \frac{2 \times CMPA + 1}{2 \times PR + 1}. \tag{6.43}$$

Due to its capability, the timer can trigger the operation of the ADC connected to the measure of $u(t)$. Indeed, the timer can trigger the ADC at every direction change of the timer, *i.e.*, $N = 0$ and $N = PR$. These times correspond to the middle of the "ON" state of K_1 and K_2, and thus, of the measure of U_d and $-U_d$. For the measurement of U_q and $-U_q$, the comparator $CMPC$ is set to $CMPC = PR/2$.

Bibliography

Apex Microtechnology, 2013. Stability for power operational amplifiers. https://www.apexanalog.com/resources/appnotes/an19u.pdf.

Arfken, G., Weber, H., Harris, F., 2013. Mathematical Methods for Physicists: A Comprehensive Guide. Elsevier Science. ISBN 978-0-12-384654-9.

Ballas, R.G., 2007. Piezoelectric Multilayer Beam Bending Actuators. Springer-Verlag, Berlin, Heidelberg.

Ben Messaoud, W, 2016. Design and Control of a Tactile Stimulator for Real Texture Simulation: Application to Textile Fabrics. Theses. Université Lille 1, Sciences et Technologies.

Bouscayrol, A., Guillaud, X., Delarue, P., Lemaire-Semail, B., 2009. Energetic macroscopic representation and inversion-based control illustrated on a wind-energy-conversion system using hardware-in-the-loop simulation. IEEE Transactions on Industrial Electronics 56 (12), 4826–4835.

Bouscayrol, A., Hautier, J.-P., Lemaire-Semail, B., 2013. Graphic Formalisms for the Control of Multi-Physical Energetic Systems: COG and EMR. John Wiley & Sons, Ltd. ISBN 9781118569863, pp. 89–124. Chapter 3.

de Abreu, G.L.C., Ribeiro, J.F., 2002. A self-organizing fuzzy logic controller for the active control of flexible structures using piezoelectric actuators. Applied Soft Computing 1 (4), 271–283.

Geradin, M., Rixen, D., 2015. Mechanical Vibrations: Theory and Application to Structural Dynamics. Wiley. ISBN 978-1-118-90020-8.

Giraud, F., Semail, B., 2006. A torque estimator for a traveling wave ultrasonic motor – application to an active claw. IEEE Transactions on Ultrasonics, Ferroelectrics, and Frequency Control 53 (8), 1468–1477.

Goldfarb, M., Celanovic, N., 1997. Modeling piezoelectric stack actuators for control of micromanipulation. IEEE Control Systems Magazine (ISSN 1066-033X) 17 (3), 69–79. https://doi.org/10.1109/37.588158.

Hemsel, T., Mracek, M., Twiefel, J., Vasiljev, P., 2006. Piezoelectric linear motor concepts based on coupling of longitudinal vibrations. Ultrasonics 44, e591–e596. Proceedings of Ultrasonics International (UI'05) and World Congress on Ultrasonics (WCU).

Ikeda, T., 1996. Fundamentals of Piezoelectricity. Oxford Science Publications. Oxford University Press. ISBN 9780198564607.

Khan, A., Giraud-Audine, C., Bigot, R., Abba, G., Abideen, M., 2016. An alternative explanation of forming force reduction for forming process submitted to vibration: influence of the waveform in the viscoplastic domain. Journal of Materials Processing Technology (ISSN 0924-0136) 230, 288–299.

Ku, S.-S., Pinsopon, U., Cetinkunt, S., Nakajima, S., 2000. Design, fabrication, and real-time neural network control of a three-degrees-of-freedom nanopositioner. IEEE/ASME Transactions on Mechatronics 5 (3), 273–280.

Ladjal, H., Hanus, J., Ferreira, A., 2009. h_∞ robustification control of existing piezoelectric-stack actuated nanomanipulators. In: 2009 IEEE International Conference on Robotics and Automation, pp. 3353–3358.

Lanczos, C., 2012. The Variational Principles of Mechanics. Dover Books on Physics. Dover Publications. ISBN 9780486134703.

Lezgy-Nazargah, M., Vidal, P., Polit, O., 2013. An efficient finite element model for static and dynamic analyses of functionally graded piezoelectric beams. Composite Structures (ISSN 0263-8223) 104, 71–84.

Li, S., Li, D., Yang, M., Cao, W., 2018. Parameters identification and contact analysis of traveling wave ultrasonic motor based on measured force and feedback voltage. Sensors and Actuators. A, Physical (ISSN 0924-4247) 284, 201–208.

Liu, X., Colli-Menchi, A.I., Gilbert, J., Friedrichs, D.A., Malang, K., Sánchez-Sinencio, E., 2015. An automatic resonance tracking scheme with maximum power transfer for piezoelectric transducers. IEEE Transactions on Industrial Electronics 62 (11), 7136–7145.

Maas, J., Schulte, T., Grotstollen, H., 1997. Optimized drive control for inverter-fed ultrasonic motors. In: Conference Record of the 1997 IEEE Industry Applications Conference Thirty-Second IAS Annual Meeting, Vol. 1. IAS '97, pp. 690–698.

Mason, W.P., 1935. An electromechanical representation of a piezoelectric crystal used as a transducer. The Bell System Technical Journal (ISSN 0005-8580) 14 (4), 718–723. https://doi.org/10.1002/j.1538-7305.1935.tb00713.x.

Maurini, C., Pouget, J., dell'Isola, F., 2006. Extension of the Euler–Bernoulli model of piezoelectric laminates to include 3d effects via a mixed approach. Computers & Structures (ISSN 0045-7949) 84 (22–23), 1438–1458. https://doi.org/10.1016/j.compstruc.2006.01.016.

Meirovitch, L., 1967. Analytical Methods in Vibrations. Macmillan Series in Advanced Mathematics and Theoretical Physics. Macmillan.

Méndez-Arriaga, F., Torres-Palma, R., Pétrier, C., Esplugas, S., Gimenez, J., Pulgarin, C., 2008. Ultrasonic treatment of water contaminated with ibuprofen. Water Research 42 (16), 4243–4248.

Mrad, R.B., Abhri, A., Zu, J., 2003. A control methodology for an inchworm piezomotor. Mechanical Systems and Signal Processing (ISSN 0888-3270) 17 (2), 457–471.

Owen, G.E., 2013. Introduction to Electromagnetic Theory. Courier Corporation. ISBN 978-0-486-17444-0.

Pan, S.H., Hudson, E.W., Davis, J.C., 1999. 3He refrigerator based very low temperature scanning tunneling microscope. Review of Scientific Instruments 70 (2), 1459–1463.

Petit, L., Gonnard, P., 2004. Inter-phases mechanical coupling in ultrasonic motors. Sensors and Actuators. A, Physical 116 (3), 492–500.

Pramanik, A., 2008. Electromagnetism: Theory and Applications. PHI Learning Pvt. Ltd. ISBN 978-81-203-3465-6.

Preumont, A., 2011. Vibration Control of Active Structures: An Introduction. Solid Mechanics and Its Applications. Springer, Netherlands. ISBN 9789400720336.

Sato, K., 2006. Trend of precision positioning technology. In: Proceedings of the ABCM Symposium Series in Mechatronics, pp. 739–750.

Sherrit, S., Leary, S.P., Dolgin, B.P., Bar-cohen, Y., 1999. Comparison of the Mason and KLM equivalent circuits for piezoelectric resonators in the thickness mode. In: IEEE Ultrasonics Symposium.

Spencer, A.J.M., 2012. Continuum Mechanics. Dover Books on Physics. Dover Publications.

Svilainis, L., Motiejunas, G., 2006. Power amplifier for ultrasonic transducer excitation. Ultragarsas "Ultrasound" (ISSN 1392-2114) 58 (1), 30–36–36. https://doi.org/10.5755/j01.u.58.1.16970.

Takasaki, M., Maruyama, Y., Mizuno, T., 2007. Resonance frequency tracing system for Langevin type ultrasonic transducers. In: 2007 International Conference on Mechatronics and Automation, pp. 3817–3822.

Tangsopa, W., Thongsri, J., 2019. Development of an industrial ultrasonic cleaning tank based on harmonic response analysis. Ultrasonics 91, 68–76.

Tiersten, H., 2013. Linear Piezoelectric Plate Vibrations: Elements of the Linear Theory of Piezoelectricity and the Vibrations Piezoelectric Plates. Springer, US. ISBN 978-1-4899-6453-3.

Uchino, K., 1997. Piezoelectric Actuators and Ultrasonic Motors. Springer, US.

Vezzoli, E., Sednaoui, T., Amberg, M., Giraud, F., Lemaire-Semail, B., 2016. Texture rendering strategies with a high fidelity – capacitive visual-haptic friction control device. In: Bello, F., Kajimoto, H., Visell, Y. (Eds.), Haptics: Perception, Devices, Control, and Applications. EuroHaptics 2016. In: Lecture Notes in Computer Science, vol. 9774. Springer, Cham.

Wallaschek, J., 1998. Contact mechanics of piezoelectric ultrasonic motors. Smart Materials and Structures 7 (3), 369.

Wikipedia contributors, 2018. Laplace transform—Wikipedia, the free encyclopedia. https://en.wikipedia.org/w/index.php?title=Laplace_transform&oldid=855860532. (Accessed 22 August 2018). Online.

Index

B

Boundary conditions, 45–46, 61–62, 64, 80
  electrical, 80

C

Closed loop, 32
Control
  CLTF, 33
  OLTF, 33
  PI, 33
  response time, 32
  static error, 32
  time constant, 33
Current
  capacitive, 87, 91, 94
  motional, 5, 87, 91, 93–94

D

D'Alembert principle, 60
Dynamic equations
  elastic, 60, 79
  piezoelectric, 45–46, 80, 82, 89

E

Elasticity
  compliance tensor, 63
  constitutive law, 59
  deformation
    displacement field, 50
  linearized strain, 52–54
  stiffness tensor, 59–60
  strain, 52, 59–60, 68
  strain components, 51, 54
  stress, 56–57, 59–60
Electric
  charge, 74–75
  crossing conditions, 77
  dielectric, 75
  dipole, 76
  equipotential, 75
  filed, 74
  flux, 74
  induction, 74, 76
  permittivity, 77–78
  polarization, 76
  potential, 74–75, 80–81, 83, 86
  susceptibility, 76
Energetic Macroscopic Representation, 99
Energy
  elastic, 60, 66, 78
  electric, 78, 83
  interaction, 78
  potential, 78

F

Frequency
  resonant, 45, 62, 64, 69, 81, 90
  response, 69–70
  spectrum, 69

M

Modal
  analysis, 45, 64, 74, 84, 86, 88
  basis, 45, 72
  coordinates, 68, 71, 83
  damping, 70
  decomposition, 45–46, 66, 69, 73, 83
  force, 68–70, 85, 87–88
  mass, 67
  truncation, 73
  velocities, 87
Mode
  normalization, 66–67, 69, 82, 90–93
  orthogonality, 66–68, 81, 83, 92
  shapes, 45–46, 62–64, 67–68, 81, 92

N

Notation
  Einstein's convention, 48–49
  Kronecker's symbol, 52
  repeated indexes, 47–48
  Van Voigt, 60

P

Partial differential equation
  eigenfunction, 62

eigenfunctions, 64–66, 83, 92
eigenvalues, 62, 64, 90
Sturm–Liouville theory, 62, 81
variables separation, 62, 64, 80
Piezoelectric
coefficient, 78
constitutive laws, 78–80
coupling factor, 79
Pull-Out Phenomena, 28, 131, 162

R

Reference configuration, 50–52, 59, 61
Rotating Reference Frame, 103, 106, 182, 192, 194

T

Theorem
Divergence, 58, 65, 68, 74, 76, 81
Expansion, 66, 88
Traveling Wave Ultrasonic Motor, 17, 123, 170

V

Vector Control Method, 42, 138, 177–178

FSC
www.fsc.org
MIX
Papier | Fördert gute Waldnutzung
FSC® C083411

Kolibri 360 GmbH
Ferdinand-Jühlke-Straße 7
99095 Erfurt, Deutschland
produktsicherheit@kolibri360.de